Lecture Notes in Artificial Intelligence 1775

Subseries of Lecture Notes in Computer Science
Edited by J. G. Carbonell and J. Siekmann

Lecture Notes in Computer Science
Edited by G. Goos, J. Hartmanis and J. van Leeuwen

Springer
Berlin
Heidelberg
New York
Barcelona
Hong Kong
London
Milan
Paris
Singapore
Tokyo

Michael Thielscher

Challenges for Action Theories

Springer

Series Editors

Jaime G. Carbonell,Carnegie Mellon University, Pittsburgh, PA, USA
Jörg Siekmann, University of Saarland, Saarbrücken, Germany

Author

Michael Thielscher
Technische Universität Dresden
Informatik, Computational Logic
01062 Dresden, Germany
E-mail: mit@inf.tu-dresden.de

Cataloging-in-Publication Data applied for

Die Deutsche Bibliothek - CIP-Einheitsaufnahme

Thielscher, Michael:
Challenges for action theories / Michael Thielscher. - Berlin ; Heidelberg ; New York ; Barcelona ; Hong Kong ; London ; Milan ; Paris ; Singapore ; Tokyo : Springer, 2000
(Lecture notes in computer science ; 1775 : Lecture notes in artificial intelligence)
ISBN 3-540-67455-1

CR Subject Classification (1991): I.2, F.4.1

ISBN 3-540-67455-1 Springer-Verlag Berlin Heidelberg New York

Springer-Verlag is a company in the BertelsmannSpringer publishing group.

Printed in Germany

Typesetting: Camera-ready by author, data conversion by PTP-Berlin, Stefan Sossna
Printed on acid-free paper SPIN 10719821 06/3142 5 4 3 2 1 0

Preface

Logic has been heralded as the basis for the next generation of computer systems. While logic and formal methods are indeed gaining grounds in many areas of computer science and artificial intelligence the expected revolution and breakthrough has not happened as yet. Notwithstanding the object oriented paradigm programming as well as processor design is still done in an imperative way, which has far-reaching consequences for the quality of software and engineering products.

A logical approach instead would offer many advantages such as machine-checked correctness, quick adaptability to design changes, dramatic reduction of maintenance costs, understandability of design, a far-reaching potential for the automation of the synthesis of the product from the design constraints, and so forth. Why then does not everyone follow the logical approach?

In the eighties is was beginning to dawn on the logic community that for most applications logic, as used then, might lack a vital ingredient which, on the other hand, is inherent in imperative languages and which no one would want to miss. What logic lacks is a simple and natural way to describe actions and change without facing inherent problems. In AI these problems center around what is called the *frame problem*. Without a solution to the frame problem—and its cousins—logic would continue to suffer from this shortcoming. Fortunately, the frame problem finally has been solved in such a way that the drawback is disappearing.

By now there is a number of formal variants of the solution to the frame problem. One consists in describing actions and change within the fluent calculus, a first-order Prolog-like formalism. The fluent calculus forms the basis of the contents of this book. It thus sets out from a basis which has overcome the drawback which held logic back for many years.

As mentioned there were the cousins of the frame problem yet to be tackled for a completely satisfactory solution. In particular, it is the so-called ramification and the qualification problem which belong to this family. Professor Thielscher in this book offers a convincing solution to these two accompanying challenges. I am deeply convinced that, as a consequence, there will be a renewed and strong interest in the logical approaches once these solutions will have become more widely known. I therefore wish this book the success it deserves for several reasons.

One of these reasons is the attractive mix of illustrative examples and formal precision, which makes this book easily accessible to a wide-spread readership. Another reason consists in the deep insight provided by the book into a fascinating topic which is central to our human thinking and has been a (mostly philosophical) issue for at least two thousand years. But the main reason is the level reached by the combined solutions to the frame problem and its cousins. Now programming in logic may comfortably include commands which call for logically defined changes without compromising in logical rigorousness. Similarly, engineering design, which always involves change, may now naturally be formalized in a logical setting with all the attractive advantages mentioned above. This includes the logical specification of agents in networks or autonomous robots which exchange information with each other as well as with human users on a most comfortable linguistic level.

It usually takes a number of years until fundamental insights diffuse through the community to a degree that the potential consequences materialize. May this book speed-up this diffusion process by finding many interested readers who spread out the news about another dawn of logic in computer science and artificial intelligence.

Wolfgang Bibel

Prologue

When John woke up he felt an uneasiness as so often these days. An instant later, however, it came to him that there was no reason for worrying anymore. The project he and his colleagues at the lab for months had sacrificed nearly everything for had finally come to a successful end the day before. So John relaxed, closed his eyes, and let his mind wander over the whole course of events again.

At the beginning there was this robot which was capable of performing rudimentary tasks such as moving around obstacles, grasping and handling objects, even climbing stairs (though it looked a bit clumsy to the attentive observer). Yet the robot was completely lacking the ability to solve tasks beyond these primitive ones on his own, that is, without John and his colleagues devising and telling him a minute plan of how to combine elementary actions in order to get the job done. A project was therefore established aiming at providing the robot both with insight into his own capabilities and with the ability to build a cognitive model of his environment. This, John argued, would enable the robot to do planning all by himself by means of reasoning when he has a certain goal in mind, that is, by drawing the right conclusions from what he knows as to the effects of his actions and from what his sensors tells him about his surroundings. A catchy name always being the basis for success of a project, they finally agreed on the acronym ELASER, meaning *Effective Logically Acting and Sensing Robot*.

It was obvious from the beginning that when explaining ELASER the effects of his actions it was impracticable to enumerate all conceivable situations in which an action can be performed and to state the result of its execution separately for each such situation. Actions had rather to be described by some sort of laws which specify the effects in general terms. In this way ELASER learnt, for instance, that grasping an object and carrying it from A to B always causes the object to be at location B and no longer at location A. In this context it turned out vital to provide the robot with a piece of information of universal nature. Namely, whenever he was told an action had such-and-such effects, he should assume this description be complete and so to conclude that moving around printed matter, for instance, does not change its contents. In order to test ELASER whether he had really grasped this crucial point, John asked him one morning to get a copy of the free local newspaper. The robot obediently walked out of the lab, spotted the right paper among different ones lying around in the front yard, correctly concluded that taking the newspaper into the building would not alter its being the local one, and so delivered it to John, who considered this an undeniable success.

Was it intuition or mere coincidence that John wanted to double-check the next morning? The night had brought some rain and it was still drizzling when ELASER left the building. Anxiously watching the robot out of his window, John saw that all newspapers were wrapped in transparent protective covers. To his great astonishment, ELASER ripped open the cover of one of the packages, tossed it away, and came back to the lab with a copy of the local newspaper soaked through. Criticizing the robot for his behavior, John and his colleagues learnt that he had no other choice: After all, ELASER explained to them, had he picked up and delivered the cover, the newspaper would still be lying in the front yard. For the single effect of carrying around an object like the cover is that only this very object changes its location, or so they had told ELASER.

Back to the drawing board. Apparently, the description of what happens if objects are moved needs to be split into two cases. Either there is no second item inside, or else there is, in which case both change their location. But what if a third object were placed inside the second one? This seems to require just another rule, which, however, still does not cover the case of four interlocking items, and so on (the alarming picture of an infinite Matryoshka, a nest of innumerable wooden puppets, entered John's mind). Pacing restlessly up and down his office racking his brain over this problem, John's eye fell on a book that had just been mailed to him with a note attached saying that it might be useful for their project. Could it be that a solution to their problem can be found in there, John thought, and so he opened the book and began to read. At least the introductory chapter seemed promising to him. The author first presented a basic theory of actions. He showed how to formally describe actions, including non-deterministic ones like rolling a dice, by specifying their general effects and applicability conditions. Furthermore, it was illustrated how to exploit this knowledge when reasoning about specific situations. The whole theory revolved around the paradigm that each action specification concentrates on what the action potentially affects, so that non-effects are to be inferred rather than being part of the description. The author called this adequacy of action specifications. So far, so good, John thought, but what if it is overly strict to suppose that nothing outside an action law is affected? Soon after starting off reading the second chapter, John realized that the latter was in fact entirely devoted to this question. There may be more to the impact of actions on the environment, so the author argued, than what is specified in action laws, which refer to the *direct* effects only. Actions may, however, have additional, *indirect* effects, which derive from general dependencies, or *constraints*, among the various properties that are used to describe the state of the environment. If John understood correctly, then this means, for instance, that an indirect effect of carrying around the cover is that the newspaper is being relocated, too. This additional effect is triggered by the general fact that two objects being stuck together can never be at different places. Accounting for indirect effects of actions, so the book continued, requires to meet two main challenges. First, the assumption needs to be suitably weakened which says that actions affect nothing but what is mentioned in action laws. Second, the aforementioned constraints often suggest, from the mere formal perspective, indirect effects which would never occur in reality and, hence, need to be sorted out. This was illustrated by an instructive example where toggling a light switch is concluded to have the magical side effect that another switch jumps its position rather than that light turns on, which one should have expected. John learnt that these two aspects together are commonly referred to as the "Ramification Problem."

With growing enthusiasm John kept on reading in hope of encountering a solution. As a matter of fact he found more than one. Unfortunately, all of them seemed perfectly reasonable to him—but only up to the point where the author proved their limited applicability. The author did so by discussing several scenarios for which the respective 'solution' either missed an obvious indirect effect (just like ELASER did in concluding that he had better rip open the cover of the newspaper, John thought), or proposed rather funny effects, which could never occur in reality. Finally, being faced with all these failures, the author introduced the concept of relations each of which directly links a single cause with a single effect. These *causal relationships* form the basis for the generation of indirect effects: Whenever a cause is brought about, then the additional occurrence of the effect is reckoned with. In this way all and also no 'phantom' indirect effects are obtained, so the author argued, provided, of course, the formal causal relationships both soundly and completely reflect causality in reality. Well, thought John, that is convincing, so we just have to provide ELASER with the knowledge that a change of an object's location *causes* any object inside to change its location as well. However, he was not looking forward to telling his colleagues that they have to draw up by hand and feed ELASER all necessary causal relationships. Fortunately the author showed how these relationships can be automatically extracted from much more general knowledge. The chapter concluded with an axiomatization in formal logic of the whole theory of causal relationships as means to solve the Ramification Problem. To John's satisfaction, he noticed that this axiomatization was based on the same principle they had used for designing ELASER—a principle which was called "Fluent Calculus" in the book. So they seem to have made a lucky decision. John remembered how difficult it was in the beginning to convince his colleagues of the advantages of this Fluent Calculus over the other's two favorites, the so-called Situation Calculus and Event Calculus, respectively, when it comes to inferring non-effects of actions. The author of the book shared his conviction, and so John and his colleagues just had to put the proposed axiomatization on top of the one already existing in ELASER.

Having redesigned ELASER following the instructions, the project members sent out the robot to fetch a newspaper each and every morning, and the robot did his duty worthily. He always came back with the right paper, plain on sunny days and safely wrapped in a protective cover whenever it had been raining. Until this one day which John will never forget. As usual, ELASER had left the lab sometime in the morning. Eventually, however, John and his colleagues realized that his return was long overdue and still there was no sign of him coming back. Anxiously recalling the disastrous morning when ELASER delivered the soaked newspaper, John followed the robot's path to the front yard. Standing next to a package with the local newspaper inside, ELASER was totally paralyzed; even when John enquired of him what had happened there was no reaction at all. John's last hope was that investigating the package lying nearby would shed some light upon the matter. Indeed he made a surprising discovery. Some rascal, who presumably had watched ELASER picking up a newspaper every morning, had teased the robot by introducing a brick into the package, which thus was too heavy for poor ELASER. Still, however, this did not account for the total blackout. When they had managed to run the robot again, he explained to John and his colleagues that he knew the only precondition of picking up an object is that it must be reachable. Now that was clearly the case when ELASER tried to lift the package, so the formal specification implied, with unerring logic, that success of this action is guaranteed. Nonetheless the expected effect failed to materialize, which entailed a logical contradiction so that ELASER's whole conception of the world broke down instantaneously. By the next morning, they had taught the robot that a second precondition for being able to lift an object is that it is sufficiently light. But then they watched ELASER anxiously ripping open the protective cover around the newspaper again, this time in order to make sure that there be no brick or any other heavy item inside. To John that seemed rather ridiculous. After all, it is highly unlikely that a newspaper package cannot be fetched on account of its weighing too much.

Back to the drawing board again. The problem was to find a suitable way of specifying action preconditions which need not be verified each time prior to assuming that the action in question be executable. John turned to the book which had already served him so well. Indeed the third chapter, entitled "The Qualification Problem," was devoted to exactly their new question. The author started off with arguing that in real-world environments most actions have many more preconditions than one is usually aware of. The reason for this unawareness is that most of these conditions are so likely to be satisfied that they are assumed away unless there is evidence to the contrary. Conditions of this sort are called "abnormal," and if their presence prevents the successful performance of an action, then the latter is considered "abnormally disqualified." Any particular situation, then, is reasonably presupposed to being 'normal' as long as this does not conflict with what is known or has been observed. John vaguely remembered this as a standard technique to deal with assumptions that are made by default because they are usually—but not necessarily always—correct. Yet the author illustrated that making the right assumptions in the context of the Qualification Problem is trickier than usual.

Roughly speaking, this is a consequence of the dynamics inherent in action theories, which implies that abnormalities may naturally arise for reasons of causality. If John understood correctly, the crucial point was the following. Suppose, for instance, ELASER had been told in advance that somebody had planned to add a brick to the newspaper package. Then it would have been reasonable for the robot to assume that this action had been successful and, hence, lifting the package would have been disqualified thereafter. But the application of the aforementioned standard technique would equally well suggest another course of events, namely, that introducing the brick is impossible in the first place due to some mysterious unspecific reason. Therefore any solution to the Qualification Problem, the author argued, requires an account of abnormal preconditions of actions which are brought about as side effects of previous actions. Side effects being nothing else than indirect effects, the preceding solution to the Ramification Problem turned out to furnish a ready fundamental for a solution to the Qualification Problem. The book continued with showing how to seek explanations in case an action surprisingly fails at some point. Even the rare case of inexplicable disqualifications, "miraculous" they were called, had been considered. Like the previous one, the chapter on the Qualification Problem concluded with an axiomatization in formal logic of the entire action theory.

Before John called an assembly of the project group in order to announce that he had found the solution to their new problem, he read through the final, comparably short chapter. The author expanded the connection between the Ramification and Qualification Problem even further. Just like actions may turn out unqualified for some abnormal reason, so the argument went, there may happen exceptions to the occurrence of indirect effects. This time the existing solution to the Qualification Problem in turn furnished a ready approach to this generalization of the Ramification Problem.

Finally, John thought contentedly daydreaming in his bed, with the invaluable assistance of the Book they had brought the project ELASER to a successful end. After having redesigned the robot so as to being capable of coping with the Qualification Problem, they continued making all kinds of tests for weeks. ELASER had passed them all effortlessly. Thinking of that, John happily nodded off again, with a hardly noticeable smile on his face. He dreamed about ELASER strolling around campus when he suddenly bumped into a famous philosopher, who straight away started arguing that the robot lacks conscious understanding of anything he is doing. When ELASER demanded a precise definition of what he meant by conscious understanding, the philosopher finally defined it as being a property only carbon-based brains can possess. I can live with having no understanding of that kind, ELASER thought walking off with a slight shake of his head, leaving the philosopher.

This, however, is not the story of ELASER. Nor is this the story of John.

This is the Book.

Table of Contents

1. Foundations of Action Theories

1.1 Purpose

Before we discuss the purpose of action theories, let us try to provide a suitable and compact definition of the subject in question:

> ***An action theory consists of a formal language that allows adequate specifications of action domains and scenarios, and it tells us precisely what conclusions can be drawn from these specifications.***

Of course this informal definition cannot be appreciated without further clarification of the crucial terms used therein. To begin with, by "action domain" we mean any aspect of the world worth formalizing in which the execution of actions plays a central role. This is the case, for instance, if one wants to model an agent that interacts with its environment, i.e., the part of the world it is able to affect. Most importantly, an *autonomous* agent needs precise knowledge as to the effects of its actions in order to act purpose-oriented and so to achieve pre-determined goals. The latter requires to draw the right conclusions from this knowledge in view of particular situations, in which the agent has acquired partial knowledge about the current state of the environment and has a certain goal in mind.

By "action scenario" we mean exactly these particular situations: Given is some information as to the current, the past and/or even a future or counterfactual state of the world. The task then is to appropriately interpret these observations so that the right conclusions can be drawn. Action theories provide this. They include a formal entailment operation that determines the set of conclusions which a scenario within an action domain allows.

Both a general action domain and a particular scenario are specified using the formal language underlying an action theory. This language determines the expressiveness of the theory. There exist action theories, for instance, that support the specification of actions with non-deterministic effects (such as rolling a dice), others don't. Whether or not a certain action theory is suitable for a particular application depends on the expressiveness required. If, for instance, it suffices to consider discrete state transitions only, then there is no need to employ an action theory designed for modeling continuous change. The two aspects discussed in this book, however, viz. indirect effects and qualifications of actions, are arguably fundamental issues and need to be

M. Thielscher: Challenges for Action Theories, LNAI 1775, pp. 1-15, 2000.

incorporated by any action theory meant to address other than artificially simplified domains.

Finally, the innocent word "adequate," which the reader may even have overlooked in the above definition, is probably the most crucial aspect. It means that the specification of actions and their effects shall be as natural as possible. For example, it would not only be inconvenient but clearly most unnatural to explicitly state the result of executing an action in every possible situation. Rather one wants to specify that, say, carrying the newspaper always causes it to change its location—no matter what color it is or whose party the current President belongs to etc. Likewise, one would want to avoid re-specifying the effects of transporting an object in case it contains (or is underneath or attached to etc.) another object. Rather the fact that this additional object also changes its location should be inferable from general knowledge of the world. In providing all this, action theories always include a more or less implicit general notion of time, change, and causality. This indicates that the adequacy requirement is what makes action theories so special—and it is what this book is all about.

Action theories have much in common with logic. They are based on a formal language and they include an entailment relation among the expressions in this language. This relation determines the way conclusions are drawn from specifications. Yet entailment in action theories is somewhat different from entailment in so-called general purpose logics, such as classical first-order logic, say. The reason is that the entailment relation reflects the special notions of change and causality inherent in action theories. This usually makes the formal definition of how to draw conclusions much more complex compared to the majority of general purpose logics. It also means that any enrichment of the ontology of an action theory necessitates changes in the definition of entailment.

Both the fair complexity and frequent changes of the notion of entailment in action theories constitute an important drawback in view of automating reasoning. This favors general purpose logics as means to this end. Research in automated deduction in first-order logic, for instance, has made noticeable progress in the past decades. It might be unwise not to exploit this development for automating reasoning in action domains. Fortunately this can be achieved without losing the major advantage of action theories, namely, their naturalness when it comes to formalizing domains and scenarios. What needs to be done towards this end is to axiomatize in, for instance, first-order logic the characteristics of an action theory. In other words, the implicit theory of change and causality is to be made explicit. It became common to call "foundational axioms" the resulting encoding that characterizes a particular action theory. Additional axioms then represent knowledge about a specific action domain and scenario. Together they provide a (hopefully) suitable encoding which allows to draw all conclusions suggested by the underlying action theory but by means of a general purpose logic. Notice that the question whether such an encoding is suitable is a precise mathematical problem and is solved by proving whether or not the drawable conclusions are always identical with what is entailed in terms of the action theory. This providing a justification for an axiomatization is a major purpose of action theories and the striving force for developing them. More to this in the brief historical account in Section 1.3.

In this book, we concern ourselves with two most fundamental challenges for action theories, namely, to account for indirect effects of actions (the Ramification Problem) and to treat unlikely but not impossible action disqualifications in a natural way (the Qualification Problem). We develop a uniform action theory that provides solutions to both problems. We further present a provably correct axiomatization of our theory by means of a general purpose logic, viz. classical first-order logic in the first part, i.e., for the Ramification Problem only, and classical logic augmented by a nonmonotonic feature for our entire theory.[1]

A Word on the Notation

The only expertise needed to understand all parts of the book is some basic knowledge of classical logic. But even this is not required except for Sections 2.9 and 3.6, where our action theory is axiomatized in formal logic.

We use the standard logical connectives, stated in order of decreasing priority, $\neg$ (negation), $\wedge$ (conjunction), $\vee$ (disjunction), $\supset$ (material implication), $\equiv$ (equivalence), $\forall$ (universal quantification), and $\exists$ (existential quantification). Both predicate symbols and constants start with a capital letter whereas function symbols and variables are in lower case, sometimes with sub- or superscripts. Free variables in formulas are assumed universally quantified unless indicated otherwise. Special symbols used in example domains are printed in typewriter style, like `turkey`, `potato`, or `tank-empty`. We further use the basic set operations and relations $\cup$ (union); $\cap$ (intersection); $\setminus$ (difference); $\in$ (membership); $\subseteq$, $\supseteq$ (sub- and superset); and $\subsetneq$, $\supsetneq$ (proper sub- and superset). All other symbols used in this book are explained at the time of their first appearance.

1.2 A Basic Action Theory

Having said much about action theories in general, we now get more formal and introduce a particular, elementary action theory containing all of the necessary basic ingredients. First of all, any action theory must provide means for describing states. A state is a snapshot of the part of the world being modeled at a particular instant of time. State descriptions need to be composed of atomic propositions. This is vital for our purposes since actions

[1] We use *Default Logic*, to be precise; see Section 3.6. The reason for pure classical logic being insufficient in view of the Qualification Problem will be given in due course.

typically affect only a small fraction of the environment—and in order to concentrate on this fraction when specifying the effects of an action we need access to it. Otherwise, i.e., if states are represented as abstract objects without bearing an internal structure (as it is typical for automata theory, for instance), the impact of an action could only be specified by a complete state transition table. This would violate the most fundamental requirement for adequacy.

Atomic propositions represent properties, or in general relations among entities, which do or do not hold in a particular state. The truth-value of any such proposition may change in the course of time as a consequence of a state transition. Due to this dynamic characteristics the term "fluent" has been established as a name for these propositions. Thus a state is characterized by saying which of the various fluents are true and which are false in this state.

Definition 1.2.1. *Let* $\mathcal{E}$ *be a finite set of symbols called* entities. *Let* $\mathcal{F}$ *denote a finite set of symbols called* fluent names, *each of which is associated with a natural number (possibly zero) called* arity. *A* fluent *is an expression* $f(e_1, \ldots, e_n)$ *where* $f \in \mathcal{F}$ *is of arity* n *and* $e_1, \ldots, e_n \in \mathcal{E}$. *A* fluent literal *is a fluent or its negation, denoted by* $\neg f(e_1, \ldots, e_n)$. *A set of fluent literals is* inconsistent *if it contains a fluent along with its negation, otherwise it is* consistent. *A* state *is a maximal consistent set of fluent literals.* ■

Before we continue, let us illustrate these concepts with a small example. At several places in this book we will model the behavior of electric circuits. Suppose a particular circuit consists of two binary switches, a light bulb, and a battery. The various states of this system may be described using the two entities $\mathcal{E} = \{\mathtt{s_1}, \mathtt{s_2}\}$ representing the two switches, along with the unary fluent `up` denoting the position of its argument, and the nullary fluent `light` denoting the state of the bulb.[2] Then both $\mathtt{up(s_2)}$ and $\neg\mathtt{light}$, say, are fluent literals, and $\{\neg\mathtt{up(s_1)}, \mathtt{up(s_2)}, \neg\mathtt{light}\}$ is a maximal consistent set of fluent literals, hence a state.

The reader may have noticed that formally any combination of truth-values constitutes a state. Later in this book, in Chapter 2, we will provide means to distinguish states that cannot occur due to domain-specific dependencies among fluents. For convenience, we introduce the following notational conventions: If ℓ is a fluent literal, then by $\|\ell\|$ we denote its affirmative component, that is, $\|f(\overline{e})\| = \|\neg f(\overline{e})\| = f(\overline{e})$ where $f \in \mathcal{F}$ and $\overline{e}$ is a sequence of n entities with n being the arity of f. This notation extends to sets of fluent literals S as follows: $\|S\| = \{\|\ell\| : \ell \in S\}$. E.g., whenever S is a state, then $\|S\|$ is the set of all fluents. If ℓ is a negative fluent literal, then $\neg\ell$ should be interpreted as $\|\ell\|$. In other words, $\neg\neg f(\overline{e}) = f(\overline{e})$. Finally, if S is a set of fluent literals, then by $\neg S$ we denote the set $\{\neg\ell : \ell \in S\}$. E.g., a set S of fluent literals is inconsistent iff $S \cap \neg S \neq \{\}$.

The second fundamental notion in any action theory are the actions themselves. Actions cause state transitions when being performed. As indicated

[2] The curious reader may take a peek at Fig. 2.3 on page 22.

above, a fundamental assumption concerning actions is that they always affect just a small fraction of an entire state. The adequacy requirement dictates that action specifications concentrate on this fraction only. Recall that the decision to split states into fluents has been made with exactly that purpose in mind. Describing the effect of an action thus amounts to specifying which fluents change their truth-value when the action is being executed. For example, switching off the second switch, $\texttt{s}_2$, always has the effect that the fluent $\texttt{up}(\texttt{s}_2)$ becomes false, regardless of the values of all other fluents. We shall formally write this as

$$\texttt{switch-off-s}_2 \ \underline{\text{transforms}} \ \{\texttt{up}(\texttt{s}_2)\} \ \underline{\text{into}} \ \{\neg\texttt{up}(\texttt{s}_2)\} \tag{1.1}$$

This means that action $\texttt{switch-off-s}_2$ can be executed in any state S which satisfies $\texttt{up}(\texttt{s}_2) \in S$, and the resulting state is obtained by replacing $\texttt{up}(\texttt{s}_2)$ by $\neg\texttt{up}(\texttt{s}_2)$ in S. So if, for instance, we switched off $\texttt{s}_2$ in the above state, $\{\neg\texttt{up}(\texttt{s}_1), \texttt{up}(\texttt{s}_2), \neg\texttt{light}\}$, then the resulting state would be $\{\neg\texttt{up}(\texttt{s}_1), \neg\texttt{up}(\texttt{s}_2), \neg\texttt{light}\}$.[3]

Specifying our example action was particularly easy because its effect is always the same. It is not hard to imagine more complex actions that produce different effects, depending on the state in which they are executed. Suppose we want to describe what happens if we *toggle* the second switch. Obviously there are two cases to be considered. First, if the switch is currently up it is down afterwards, and, second, if it is already down in the beginning, then toggling it causes it to be in the upper position. Actions of this more complex kind are described by two or more "laws" of the form of (1.1). Toggling the second switch, for instance, can be specified through

$$\begin{array}{llll} \texttt{toggle-s}_2 & \underline{\text{transforms}} & \{\texttt{up}(\texttt{s}_2)\} & \underline{\text{into}} & \{\neg\texttt{up}(\texttt{s}_2)\} \\ \texttt{toggle-s}_2 & \underline{\text{transforms}} & \{\neg\texttt{up}(\texttt{s}_2)\} & \underline{\text{into}} & \{\texttt{up}(\texttt{s}_2)\} \end{array}$$

Notice that since states are maximal consistent sets, they always contain either $\texttt{up}(\texttt{s}_2)$ or else $\neg\texttt{up}(\texttt{s}_2)$, at least in our example domain. In any case, therefore, exactly one of the two action laws for $\texttt{toggle-s}_2$ is applicable and yields the expected result, namely, that the switch alters its position.

It is often convenient to specify an action simultaneously for a whole collection of similar entities. To this end, action laws may contain variables to be substituted by entities. For instance, the two specifications

$$\begin{array}{lllll} \texttt{toggle}(x) & \underline{\text{transforms}} & \{\texttt{up}(x)\} & \underline{\text{into}} & \{\neg\texttt{up}(x)\} \\ \texttt{toggle}(x) & \underline{\text{transforms}} & \{\neg\texttt{up}(x)\} & \underline{\text{into}} & \{\texttt{up}(x)\} \end{array} \tag{1.2}$$

describe what toggling means in general. To be more precise, let $\overline{x}$ denote a finite sequence of variables chosen from a given denumerable set $\mathcal{V}$. If $\overline{x}$ contains the variables that occur in some expression ξ, then this is written

[3] The reader is right who wonders whether this really is all that is to be said as to the effect of turning off a switch. More to this in Chapter 2.

$\xi[\overline{x}]$. Let $\overline{x} = x_1, \ldots, x_n$, then a *ground instance* of an expression $\xi[\overline{x}]$ is obtained by applying a substitution $\theta = \{x_1 \mapsto e_1, \ldots, x_n \mapsto e_n\}$ to ξ, where $e_1, \ldots, e_n \in \mathcal{E}$ are entities.[4] Suppose $\overline{e} = e_1, \ldots, e_n$, then $\xi[\overline{x}]\theta$ is also written $\xi[\overline{e}]$. For example, the ground instance

$$\alpha[\mathtt{s_2}] = \mathtt{toggle}(\mathtt{s_2}) \ \underline{\text{transforms}} \ \{\mathtt{up}(\mathtt{s_2})\} \ \underline{\text{into}} \ \{\neg\mathtt{up}(\mathtt{s_2})\}$$

can be obtained from $\alpha[x] = \mathtt{toggle}(x)$ $\underline{\text{transforms}}$ $\{\mathtt{up}(x)\}$ $\underline{\text{into}}$ $\{\neg\mathtt{up}(x)\}$ by substituting $\{x \mapsto \mathtt{s_2}\}$.[5]

This finally leads us to the formal definition of actions and their specification. We call *fluent expressions* any $f(t_1, \ldots, t_n)$ and its negation $\neg f(t_1, \ldots, t_n)$ where f is a fluent name of arity $n \geq 0$ and each t_i is either an entity or a variable $(1 \leq i \leq n)$.

Definition 1.2.2. *Let $\mathcal{A}$ be a finite set of action names, each of which is associated with a natural number (possibly zero) called* arity. *An* action *is a ground term $a(\overline{e})$ where $a \in \mathcal{A}$ is of arity equal to the length of $\overline{e}$. An* action law *is of the form*

$$a(\overline{x}) \ \underline{\text{transforms}} \ C[\overline{x}] \ \underline{\text{into}} \ E[\overline{x}]$$

where $a \in \mathcal{A}$ is of arity equal to the length of $\overline{x}$ and where $C[\overline{x}]$ and $E[\overline{x}]$ are sets of fluent expressions which satisfy the following. Both $C[\overline{e}]$ and $E[\overline{e}]$, for any $\overline{e}$, are consistent and $\|C[\overline{e}]\| = \|E[\overline{e}]\|$, that is, C and E always refer to the same fluents. If S is a state, then a ground instance $\alpha[\overline{e}]$ of an action law $\alpha[\overline{x}] = a(\overline{x})$ $\underline{\text{transforms}}$ $C[\overline{x}]$ $\underline{\text{into}}$ $E[\overline{x}]$ is applicable *to S iff $C[\overline{e}] \subseteq S$. The* application *of $\alpha[\overline{e}]$ to S yields $(S \setminus C[\overline{e}]) \cup E[\overline{e}]$; the latter is called* preliminary successor state *of S and $a(\overline{e})$.*[6] ∎

The reader may notice that S being a state, $C[\overline{e}]$ and $E[\overline{e}]$ being consistent, and $\|C[\overline{e}]\| = \|E[\overline{e}]\|$ guarantee that $(S \setminus C[\overline{e}]) \cup E[\overline{e}]$ is a state, too.

Thus far we have only seen actions which determine at most one preliminary successor state when being executed. We needed two different action laws to specify the effects of $\mathtt{toggle}(x)$, that much is true, but the two conditions were mutually exclusive so that the two laws are never applicable at the same time. Yet Definition 1.2.2 does not exclude the existence of two (or more) simultaneously applicable laws for one and the same action. This feature is required if one wants to describe actions with indeterminate effects, so-called non-deterministic actions. Tossing a coin is a simple action which

---

[4] Application of a substitution $\theta = \{x_1 \mapsto e_1, \ldots, x_n \mapsto e_n\}$ to an expression ξ means the replacement, in ξ, of any occurrence of variable x_i by entity e_i $(1 \leq i \leq n)$. The resulting expression is denoted by $\xi\theta$.

[5] Had we considered entities different from switches and for which the notion of toggling would make no sense, such as, say, potatoes (see Chapter 3), then we should introduce a unary fluent `switch` denoting whether or not its argument is a switch. This fluent should then be used as an additional condition within the two action laws for $\mathtt{toggle}(x)$.

[6] The reason for calling "preliminary" this state will be revealed in Chapter 2.

obviously belongs to that category. But to stick to our current domain, suppose it is totally dark so that it is impossible to tell the two switches apart. Nonetheless we want to switch up one of them (knowing both are currently down). Putting this plan into execution, there are two possible outcomes: We either hit the first or else the second switch. This may be formalized by the two action laws

$$\begin{array}{llll} \texttt{switch-one-up} & \underline{\text{transforms}} & \{\neg\texttt{up}(\texttt{s}_1)\} & \underline{\text{into}} \ \{\texttt{up}(\texttt{s}_1)\} \\ \texttt{switch-one-up} & \underline{\text{transforms}} & \{\neg\texttt{up}(\texttt{s}_2)\} & \underline{\text{into}} \ \{\texttt{up}(\texttt{s}_2)\} \end{array} \tag{1.3}$$

According to Definition 1.2.2 these two laws are simultaneously applicable to the state $\{\neg\texttt{up}(\texttt{s}_1), \neg\texttt{up}(\texttt{s}_2), \neg\texttt{light}\}$. Their respective application determines two different successor states, viz. $\{\texttt{up}(\texttt{s}_1), \neg\texttt{up}(\texttt{s}_2), \neg\texttt{light}\}$ and $\{\neg\texttt{up}(\texttt{s}_1), \texttt{up}(\texttt{s}_2), \neg\texttt{light}\}$. In any *particular* situation of course only one of these possibilities will actually occur. Which one this will be, however, cannot be predicted, at least not on the basis of the (restricted) knowledge about the domain. This is what makes the action in question non-deterministic.[7]

The concept of action laws defining how the execution of actions affects the particles of state descriptions, viz. the fluents, provides us with a basic formalism to specify action domains. The semantics of these specifications is given by complete state transition models.

Definition 1.2.3. *A* basic action domain $\mathcal{D}$ *is a 4-tuple* $(\mathcal{E}, \mathcal{F}, \mathcal{A}, \mathcal{L})$ *where* $\mathcal{E}$ *is a set of entities,* $\mathcal{F}$ *a set of fluent names,* $\mathcal{A}$ *a set of action names, and* $\mathcal{L}$ *is a set of action laws. The* transition model *of* $\mathcal{D}$ *is a total mapping* Σ *from state-action pairs into (possibly empty) sets of states such that* $S' \in \Sigma(S, a)$ *iff* S' *is a preliminary successor of* S *and* a*.* ■

Any (basic) action domain provides general, i.e., situation-independent knowledge as to the impact of performing actions. Exploiting this knowledge when drawing conclusions about particular scenarios within a domain is our next concern.

A scenario is given by information as to particular developments of the part of the world which has been specified as action domain. In most cases, this information is incomplete in that only for some fluents at some stages the truth-values are known. The general task then is to draw the right conclusions as to the truth-values of other fluents at other stages. Of course this requires knowledge of the general effects of actions, given by the formal specification of the underlying action domain.

As a very simple example, suppose we observe that after switch $\texttt{s}_1$ has been toggled it is in the upper position. Then it is reasonable to conclude that the switch was down beforehand. This should follow from our knowledge as to the general effects of toggling switches, which is provided by the two action laws defined in (1.2). Formally, we will use expressions like

[7] Although non-determinism is not amongst the basic requirements for action theories, it will be vital for both the Ramification and Qualification Problem.

$\texttt{up}(\texttt{s}_1)$ <u>after</u> $[\texttt{toggle}(\texttt{s}_1)]$ to denote what we call *observations*. A scenario can then be modeled as a collection of observations, and drawing the right conclusions amounts to deciding which observations follow from the given ones. E.g., $\neg\texttt{up}(\texttt{s}_1)$ <u>after</u> $[\,]$ could be one such conclusion.

Observations may be more complex than the two examples just mentioned. For instance, $\texttt{up}(\texttt{s}_1) \equiv \texttt{up}(\texttt{s}_2)$ <u>after</u> $[\,]$ says that both switches are in the same position—though it is not known what position they share. A reasonable conclusion here would be, say, $\neg(\texttt{up}(\texttt{s}_1) \equiv \texttt{up}(\texttt{s}_2))$ <u>after</u> $[\texttt{toggle}(\texttt{s}_2)]$, that is, toggling the second switch results in both switches being in different positions. Another variant is to ask hypothetical questions like the following. Suppose the result of toggling $\texttt{s}_2$ would be that all switches have the same position. Suppose further that toggling $\texttt{s}_1$ would bring this switch down. What, then, follows as for the result of toggling both $\texttt{s}_1$ and $\texttt{s}_2$? Clearly, the assumptions imply that initially $\texttt{s}_1$ is up and (hence) $\texttt{s}_2$ is down. From this we conclude that the right answer to the question is that switch $\texttt{s}_1$ is down and switch $\texttt{s}_2$ is up. More formally, we say that the two observations

$$\begin{array}{r} \forall x.\, \texttt{up}(x) \vee \forall x.\, \neg\texttt{up}(x) \ \underline{\text{after}}\ [\texttt{toggle}(\texttt{s}_2)] \\ \neg\texttt{up}(\texttt{s}_1) \ \underline{\text{after}}\ [\texttt{toggle}(\texttt{s}_1)] \end{array} \tag{1.4}$$

entail the observation $\neg\texttt{up}(\texttt{s}_1) \wedge \texttt{up}(\texttt{s}_2)$ <u>after</u> $[\texttt{toggle}(\texttt{s}_1), \texttt{toggle}(\texttt{s}_2)]$. Again, to reiterate the obvious, the conclusion relies on the correct definition of the action laws. In the remainder of this section, we show how conclusions of this kind are obtained on a formal basis. To this end, we first need a precise definition of observations. This in turn requires the formal notion of a so-called fluent formula, such as $\neg(\texttt{up}(\texttt{s}_1) \equiv \texttt{up}(\texttt{s}_2))$ or $\forall x.\, \neg\texttt{up}(x)$.

Definition 1.2.4. *Given sets of entities, fluent names, and variables, the set of* fluent formulas *is inductively defined as follows: Each fluent expression and* $\top$ *(tautology) and* $\bot$ *(contradiction) are fluent formulas, and if* F *and* G *are fluent formulas so are* $\neg F$, $F \wedge G$, $F \vee G$, $F \supset G$, $F \equiv G$, $\exists x.\, F$, *and* $\forall x.\, F$ *(where* x *is a variable). A* closed *formula is a fluent formula without free variables, that is, where variables only occur in the scope of some quantifier using this variable.*[8] ■

Definition 1.2.5. *Let* $\mathcal{E}$, $\mathcal{F}$, *and* $\mathcal{A}$ *be sets of entities, fluent names, and action names, respectively. An* observation *is of the form*

$$F \ \underline{\text{after}}\ [a_1(\overline{e}_1), \ldots, a_n(\overline{e}_n)]$$

where F *is a closed fluent formula and each of* $a_1(\overline{e}_1), \ldots, a_n(\overline{e}_n)$ *is an action* $(n \geq 0)$. ■

Definition 1.2.6. *A* basic action scenario *is a pair* $(\mathcal{O}, \mathcal{D})$ *where* $\mathcal{D}$ *is a basic action domain and* $\mathcal{O}$ *is a set of observations (based on the entities and fluent and action names in* $\mathcal{D}$*).* ■

[8] The *scope* of the quantifiers in $\exists x.\, F$ and $\forall x.\, F$ is defined as the subformula F.

Prior to defining what conclusions an action scenario allows, we need to clarify under which circumstances a particular observation can be said to be true. This obviously depends on what state supposedly results from executing the action sequence in question. If that state is fixed, then deciding whether the fluent formula itself holds is straightforward, following the standard interpretation of the logical connectives.

Definition 1.2.7. *Let $\mathcal{E}$ and $\mathcal{F}$ be sets of entities and fluent names, respectively, and let S be a state. The notion of a closed formula being* true *(resp.* false*) in S is inductively defined as follows:*

1. $\top$ *is true and* $\bot$ *is false in S;*
2. *a fluent literal ℓ is true in S iff $\ell \in S$;*
3. $\neg F$ *is true in S iff F is false in S;*
4. $F \wedge G$ *is true in S iff F and G are true in S;*
5. $F \vee G$ *is true in S iff either F or G is true in S (or both);*
6. $F \supset G$ *is true in S iff F is false in S or G is true in S (or both);*
7. $F \equiv G$ *is true in S iff F and G are true in S, or else F and G are false in S;*
8. $\exists x. F$ *is true in S iff there exists some $e \in \mathcal{E}$ such that $F\{x \mapsto e\}$ is true in S;*
9. $\forall x. F$ *is true in S iff for each $e \in \mathcal{E}$, $F\{x \mapsto e\}$ is true in S.*

Here, $F\{x \mapsto e\}$ denotes the fluent formula resulting from replacing in F all free occurrences of variable x by entity e. ■

As an example consider the formula $\exists x. \neg \mathtt{up}(x) \supset \neg \mathtt{light}$, which is true in the state $\{\neg \mathtt{up}(\mathtt{s}_1), \mathtt{up}(\mathtt{s}_2), \neg \mathtt{light}\}$ (since $\neg \mathtt{light}$ is true) and also in $\{\mathtt{up}(\mathtt{s}_1), \mathtt{up}(\mathtt{s}_2), \mathtt{light}\}$ (since $\exists x. \neg \mathtt{up}(x)$ is false), but the formula is false in, e.g., $\{\mathtt{up}(\mathtt{s}_1), \neg \mathtt{up}(\mathtt{s}_2), \mathtt{light}\}$.

As indicated, the observations that describe a scenario usually provide only incomplete information as to the entire state of affairs. This is especially true if non-deterministic actions are considered because then complete information means to know the actual result of any possible sequence of non-deterministic actions. Thus the normal case is that there is more than just one unique state of affairs that fits a scenario description. Following a standard terminology in logic, we call any possible state of affairs an *interpretation*, and if the latter matches a scenario description it is called a *model* thereof.

We have already seen examples involving reasoning about hypothetical developments of the world. An interpretation therefore must not just tell us exactly what happens during the execution of one particular sequence of actions. Rather it needs to provide this information as to any possible course of events. Of course we assume the world always evolves according to the underlying action laws. That is to say, whenever some state S results from performing some action sequence, and some further action a is executed, then the result should be a successor of S and a.

Definition 1.2.8. *Let* $(\mathcal{O},\mathcal{D})$ *be a basic action scenario. An* interpretation *for* $(\mathcal{O},\mathcal{D})$ *is a pair* (Σ, Res) *where* Σ *is the transition model of* $\mathcal{D}$ *and* Res *is a partial function which maps finite (possibly empty) action sequences to states and which satisfies the following:*

1. $Res([\,])$ *is defined.*
2. *For any sequence* $a^* = [a_1, \ldots, a_{k-1}, a_k]$ *of actions* $(k > 0)$*,*
   a) $Res(a^*)$ *is defined if and only if* $Res([a_1, \ldots, a_{k-1}])$ *is defined and* $\Sigma(Res([a_1, \ldots, a_{k-1}]), a_k)$ *is not empty, and*
   b) $Res(a^*) \in \Sigma(Res([a_1, \ldots, a_{k-1}]), a_k)$. ■

The second component of an interpretation may be depicted as a tree (of infinite depth) whose root node contains the initial state, $Res([\,])$, and whose branches each characterize the supposed evolution of the world under a particular sequence of actions; see Fig. 1.1.

Interpretations always tell us the exact result of performing any possible action sequence. It is therefore straightforward to determine whether an observation is true with regard to a particular interpretation: First of all, it can be true only if the state is defined which results from performing the sequence of actions in question. If, moreover, the fluent formula in question is true in that state, then the observation itself is true.[9]

Definition 1.2.9. *Let* (Σ, Res) *be an interpretation for a basic action scenario* $(\mathcal{O},\mathcal{D})$*. An observation* F after $[a_1, \ldots, a_n]$ $(n \geq 0)$ *is* true *in* Res *iff* $Res([a_1, \ldots, a_n])$ *is defined and* F *is true in* $Res([a_1, \ldots, a_n])$. ■

For the purpose of illustration, the reader may verify that the two observations (1.4) from above are true in the interpretation depicted in Fig. 1.1, but not the observation `up(s1)` after `[toggle(s1), switch-one-up]`, nor the observation `¬light` after `[toggle(s2), switch-one-up]` because the result of the latter action sequence is undefined.

Among all possible interpretations for the underlying action domain we are especially interested in those which satisfy all observations of a specific scenario. As indicated, these are called the models of the scenario. Models help us define what can be concluded from a scenario description, namely, any observation which is true in *all* models of a domain.

Definition 1.2.10. *Let* $(\mathcal{O},\mathcal{D})$ *be a basic action scenario. A* model *of this scenario is an interpretation* (Σ, Res) *such that each* $o \in \mathcal{O}$ *is true in* Res*. An observation* o *is* entailed*, written* $\mathcal{O} \models_{\mathcal{D}} o$*, iff* o *is true in all models of* $(\mathcal{O},\mathcal{D})$. ■

This completes the introduction to our basic action theory. Now that we have put together all necessary formal concepts, let us give a fully formalized

[9] An observation F after a^* is considered false in an interpretation Res whenever $Res(a^*)$ is undefined (the alternative would be to consider it undefined, too), because we take the observation as implicitly asserting that the action sequence a^* be executable.

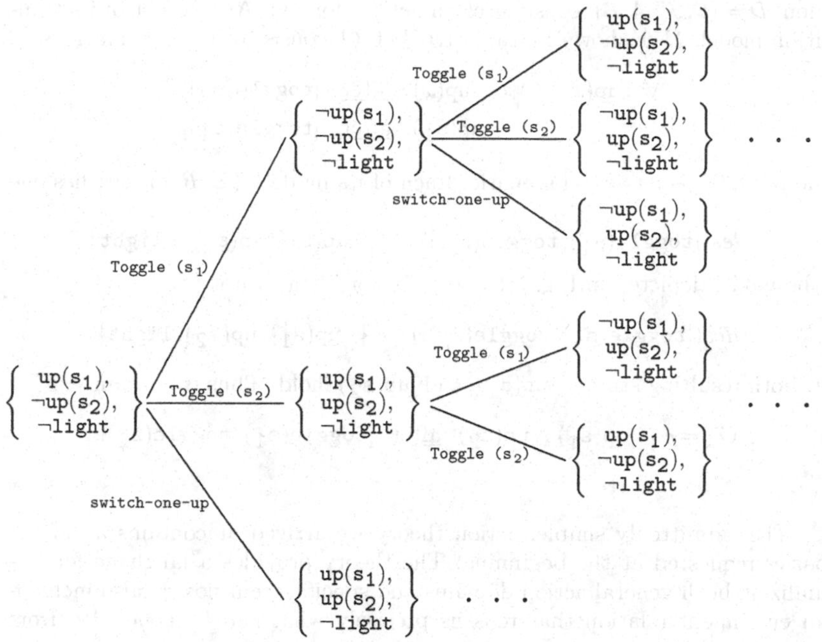

Figure 1.1. Each interpretation suggests possible evolutions of the system's state. This may be depicted as tree whose nodes are states and whose edges are labeled with actions. In the example, the state resulting from performing $\texttt{toggle}(\texttt{s}_2)$ admits only two child nodes, as opposed to, say, the root node admitting three of them. The reason is that no action law for **switch-one-up** (c.f. (1.3)) is applicable to the state $\{\texttt{up}(\texttt{s}_1), \texttt{up}(\texttt{s}_2), \neg\texttt{light}\}$. Thus *Res*([$\texttt{toggle}(\texttt{s}_2)$, **switch-one-up**]) is not defined in this interpretation. Notice further that when constructing an interpretation one has to select among the alternative results whenever a non-deterministic action is performed; in the example the result of [$\texttt{toggle}(\texttt{s}_1)$, **switch-one-up**], for instance, has been determined as $\{\neg\texttt{up}(\texttt{s}_1), \texttt{up}(\texttt{s}_2), \neg\texttt{light}\}$—we would have a different interpretation had we made the equally possible choice $\{\texttt{up}(\texttt{s}_1), \neg\texttt{up}(\texttt{s}_2), \neg\texttt{light}\}$.

version of our running example. Below and from now onwards, by ξ^k we indicate that ξ is of arity k.

Example 1.2.1. Let $\mathcal{E} = \{\texttt{s}_1, \texttt{s}_2\}$ be the set of entities, $\mathcal{F} = \{\texttt{up}^1, \texttt{light}^0\}$ the set of fluent names, and $\mathcal{A} = \{\texttt{toggle}^1, \texttt{switch-one-up}^0\}$ the set of action names. Furthermore, let $\mathcal{L}$ consist of the action laws

$$\begin{array}{rlll}
\texttt{toggle}(x) & \underline{\text{transforms}} & \{\texttt{up}(x)\} & \underline{\text{into}} & \{\neg\texttt{up}(x)\} \\
\texttt{toggle}(x) & \underline{\text{transforms}} & \{\neg\texttt{up}(x)\} & \underline{\text{into}} & \{\texttt{up}(x)\} \\
\texttt{switch-one-up} & \underline{\text{transforms}} & \{\neg\texttt{up}(\texttt{s}_1)\} & \underline{\text{into}} & \{\texttt{up}(\texttt{s}_1)\} \\
\texttt{switch-one-up} & \underline{\text{transforms}} & \{\neg\texttt{up}(\texttt{s}_2)\} & \underline{\text{into}} & \{\texttt{up}(\texttt{s}_2)\}
\end{array}$$

then $\mathcal{D} = (\mathcal{E}, \mathcal{F}, \mathcal{A}, \mathcal{L})$ constitutes an action domain. An excerpt of its transition model Σ is shown in Table 1.1. Let $\mathcal{O}$ consist of the observations

$$\forall x.\, \mathtt{up}(x) \vee \forall x.\, \neg\mathtt{up}(x) \;\; \underline{\text{after}} \;\; [\mathtt{toggle(s_2)}]$$
$$\neg\mathtt{up(s_1)} \;\; \underline{\text{after}} \;\; [\mathtt{toggle(s_1)}]$$

then $(\mathcal{O}, \mathcal{D})$ is an action scenario. Each of its models (Σ, Res) satisfies one of

$$Res([\mathtt{toggle(s_1)}, \mathtt{toggle(s_2)}]) = \{\neg\mathtt{up(s_1)}, \mathtt{up(s_2)}, \neg\mathtt{light}\}$$

(the model depicted in Fig. 1.1 does this, for instance) or

$$Res([\mathtt{toggle(s_1)}, \mathtt{toggle(s_2)}]) = \{\neg\mathtt{up(s_1)}, \mathtt{up(s_2)}, \mathtt{light}\}$$

In both resulting states $\neg\mathtt{up(s_1)}$ and $\mathtt{up(s_2)}$ hold. Thus we obtain

$$\mathcal{O} \models_{\mathcal{D}} \neg\mathtt{up(s_1)} \wedge \mathtt{up(s_2)} \;\; \underline{\text{after}} \;\; [\mathtt{toggle(s_1)}, \mathtt{toggle(s_2)}]$$

■

The, admittedly simple, action theory we arrived at contains all of the basics requested at the beginning: The theory provides a language for formalizing both general action domains and specific scenarios. It also includes an entailment relation that tells us precisely what can be concluded from

Table 1.1. The transition model assigns each pair of state and action a set of preliminary successor states.

S	a	$\Sigma(S,a)$
$\{\neg\mathtt{up(s_1)}, \neg\mathtt{up(s_2)}, \neg\mathtt{light}\}$	$\mathtt{toggle(s_1)}$	$\{\{\mathtt{up(s_1)}, \neg\mathtt{up(s_2)}, \neg\mathtt{light}\}\}$
	$\mathtt{toggle(s_2)}$	$\{\{\neg\mathtt{up(s_1)}, \mathtt{up(s_2)}, \neg\mathtt{light}\}\}$
	$\mathtt{switch\text{-}one\text{-}up}$	$\left\{\begin{array}{l}\{\mathtt{up(s_1)}, \neg\mathtt{up(s_2)}, \neg\mathtt{light}\},\\ \{\neg\mathtt{up(s_1)}, \mathtt{up(s_2)}, \neg\mathtt{light}\}\end{array}\right\}$
$\{\mathtt{up(s_1)}, \neg\mathtt{up(s_2)}, \neg\mathtt{light}\}$	$\mathtt{toggle(s_1)}$	$\{\{\neg\mathtt{up(s_1)}, \neg\mathtt{up(s_2)}, \neg\mathtt{light}\}\}$
	$\mathtt{toggle(s_2)}$	$\{\{\mathtt{up(s_1)}, \mathtt{up(s_2)}, \neg\mathtt{light}\}\}$
	$\mathtt{switch\text{-}one\text{-}up}$	$\{\{\mathtt{up(s_1)}, \mathtt{up(s_2)}, \neg\mathtt{light}\}\}$
⋮	⋮	⋮
$\{\mathtt{up(s_1)}, \mathtt{up(s_2)}, \mathtt{light}\}$	$\mathtt{toggle(s_1)}$	$\{\{\neg\mathtt{up(s_1)}, \mathtt{up(s_2)}, \mathtt{light}\}\}$
	$\mathtt{toggle(s_2)}$	$\{\{\mathtt{up(s_1)}, \neg\mathtt{up(s_2)}, \mathtt{light}\}\}$
	$\mathtt{switch\text{-}one\text{-}up}$	$\{\}$

a scenario description. Our theory satisfies the most fundamental adequacy requirement, namely, that action laws concentrate on the part of the world which is affected when the action is performed. On the other hand, the basic theory makes two very strong assumptions. First, it requires any action law to contain the entire effect of the action it describes, that is, not only the direct but also all possible indirect effects. This is a consequence of the assumption that no fluent changes which is not mentioned in an action law. Second, the theory asserts that an action is guaranteed to succeed once all of the specified conditions are true in a state. Both assumptions, their inadequacy, and a much more sophisticated action theory which overcomes these assumptions are subject of the remaining chapters.

1.3 Bibliographic Remarks

Formal action theories have not been developed until very recently. The underlying idea of formalizing reasoning about actions and planning, however, is much older. In fact, automating the ability of common sense to reason about actions and their effects was among the very first issues raised in Artificial Intelligence research [75]. There the belief was advocated that multifarious intelligent behavior relies on the ability to maintain a mental model of the world and to draw the right conclusions about observations and intentions.

The historically first formal approach to reasoning about actions was a pure first-order encoding of some example action domains and scenarios. This encoding introduced the so-called *Situation Calculus* paradigm, which satisfies the fundamental requirement for adequacy in that actions are specified by their effects. In so doing, the Situation Calculus brings along the *Frame Problem* [74], which denotes the problem both of how to represent, in logic, the general assumption that any non-affected fluent keeps its truth-value when an action is performed, and of how to reason efficiently with this representation.[10]

While the earliest Situation Calculus-based encodings were intuitively plausible, the solutions to the Frame Problem suggested and employed in this context were cumbersome and inefficient [43]. Most of subsequent work in this field was therefore devoted to finding better solutions to this problem. It is beyond all question that conceivable progress had been made in this regard,[11] but it turned out that a considerable price has been paid towards this end: Axiomatizations of action scenarios seem to become lesser

[10] Philosophers have dealt with problems related to actions and their effects for much longer, in the context of causality. Interestingly enough, however, they seem never to have struggled with the adequacy of action specifications. This is indicated by the fact that the Frame Problem has not been encountered until AI researchers started to investigate action formalisms [22]. Even later formal approaches to the causality phenomenon, such as [120, 1], presuppose exhaustive state transition functions.

[11] more to this in Section 2.10

plausible the better they tackle the Frame Problem. Triggered by an increasing number of action formalisms being erroneous in allowing unintended and counter-intuitive conclusions, the reliability of action encodings became a new major problem. The by far most popular case is the approach proposed in [78], which uses so-called circumscription [77], a particular extension to classical first-order logic, to address the Frame Problem. Following common practice, this approach was validated merely by example. Soon, however, a serious flaw of this method was revealed with the help of yet another simple example[12] which is treated against the intuition [49].[13]

The increasing difficulties with assessing the correctness of action formalisms just by appealing to the intuition led to the insight that formal validation methods are needed in order to guarantee reliability. The first step in this direction was a generalization of an intuitively correct encoding of an example domain so that a whole, well-defined class of domains is covered [66]. This generalization was accompanied by a formal proof that the encoding of any such domain will yield the same results as obtained in the example, hence will inherit the intuitive correctness. Although this result was established for a very restricted class of domains (compared, e.g., to what is expressible in our basic action theory), this was the first time that an action formalism itself was subject to a formal proof of its correctness.

The first action theory that deserves this name in being truly independent of a specific axiomatization was the *Action Description Language*, abbreviated $\mathcal{A}$ [34].[14] This formalism and our basic action theory have similar expressiveness except that $\mathcal{A}$ is restricted to nullary (i.e., propositional) fluents and does not support non-deterministic actions. In fact, many of our notions and notations were borrowed from this first approach. The Action Description Language was first employed to validate an encoding of action domains based on Situation Calculus and using so-called extended logic programs [34]. Later, $\mathcal{A}$ has been employed for the evaluation of action formalisms such as Situation Calculus based on abductive logic programs [24, 20], Situation Calculus with so-called successor state axioms (see Section 2.10) and Situation Calculus with circumscription, respectively [58], and Fluent Calculus (see Section 2.9) based on so-called equational logic programs [108], just to mention a few. Any of these results verifies correctness of the respective action formalisms if applied to domains expressible in $\mathcal{A}$. In order to obtain more general assessment results, the original Action Description Language has been extended into various directions, among which we again mention just a few.

[12] This famous example is known as the *Yale Shooting* scenario; see also Sections 3.7 and 2.8.

[13] Moreover, the improvement to the original formulation proposed in [3], again motivated solely by examples, has been proved (by a counter-example) restricted to a stronger extent than originally claimed [59].

[14] First published as [33]. The reader should not be misled by the name; $\mathcal{A}$ is not just a language but comes along with a formal semantics of how expressions in this language are to be interpreted in terms of state transitions, models, and entailment.

Non-propositional fluents were first introduced in [24]. The dialect $\mathcal{A}_C$ allows to represent and reason about the effects of concurrently performed actions [4]. Non-deterministic actions are supported in $\mathcal{A}_N$, and the language $\mathcal{A}_{NCC}$ extends the latter and combines it with $\mathcal{A}_C$ [12]. In $\mathcal{AR}$, too, non-deterministic actions can be formalized and, moreover, it includes a basic solution to the problem of indirect effects of actions [57, 40]. The extension $\mathcal{L}_0$ supports incomplete specification of the action sequence that actually occurs in a scenario [6]. Information gathering actions can be represented in the language $\mathcal{A}_k$ [70]. The language $\mathcal{E}$ of [55] is based on a linear time structure and allows to represent narratives of events. A good general introduction to the Action Description Language and several dialects is [35].

The second major class of action theories has its roots in the language and semantics developed in [89]. A distinctive feature of this framework, which is commonly called "Features-and-Fluents," is that it allows to reason about the duration of actions. A thorough and fine-grained hierarchy of sub-classes with restricted expressiveness allows precise assessment results for different action formalisms. Major evaluation results are reported in [90]. Several extensions to the original framework have been developed, e.g., concurrency of actions [122], actions with indirect effects [91, 47, 93], and goal-orientedness [95]. The reference article [94] offers a general introduction to this line of research.

A first formal comparison between Action Description Language $\mathcal{A}$ and "Features-and-Fluents" was established in [107], where both theories are proved equivalent for a particular class of domains. Of course many more formalisms exist which arguably deserve being called action theories. The Action Description Language and the "Features-and-Fluents" framework, however, are the only ones that found broader dissemination in two respects: They have been employed for the validation of a variety of action formalisms, and they have been used as a uniform basis for dealing with a variety of ontological aspects. Insofar as these and other more specialized action theories address the problems we are concerned with in the following chapters, they will be described and discussed in detail in due course.

Nonmonotonic fluents were first introduced in [2]. The [illegible] [illegible] [illegible] and [illegible] about [illegible] effects of nonmonotony [illegible] [illegible]. [illegible] are supported in [illegible] and [illegible] [illegible] [illegible] [illegible] [illegible] [12] [illegible] [illegible] [illegible] [illegible] [illegible] [illegible] [illegible] [illegible] to the problem of indirect effects of actions [illegible]. The [illegible] [illegible] [illegible] [illegible] [illegible] [illegible] [illegible] [illegible] [illegible] [illegible] language [illegible] [illegible] [illegible] [illegible] time [illegible] and [illegible] [illegible] [illegible] [illegible] introduced to the Action Description Language and several [illegible] [illegible].

The second [illegible] [illegible] [illegible] [illegible] [illegible] [illegible] to the language [illegible] [illegible] [illegible] [illegible] [illegible] [illegible] which is [illegible] [illegible] [illegible] and [illegible] is that it allows to [illegible] [illegible] [illegible] [illegible]. A thorough and [illegible] hierarchy of [illegible] [illegible] and [illegible] [illegible] [illegible] [illegible] [illegible] [illegible] [illegible] [illegible] [illegible] reported in [illegible]. Several [illegible] to the [illegible] [illegible] [illegible] [illegible] [illegible] [illegible] [illegible] [illegible] [22] [illegible] [illegible] [illegible] [illegible] [illegible] [illegible] [illegible] [illegible] [illegible] [illegible] [illegible] [illegible] [illegible].

A [illegible] [illegible] [illegible] between Action Description Languages [illegible] [illegible] [illegible] [illegible] [illegible] [illegible] [illegible] [illegible] [illegible] [illegible] [illegible] [illegible] [illegible] class of domains [illegible] [illegible] [illegible] [illegible] [illegible] the [illegible] called [illegible] [illegible] Action [illegible] [illegible] [illegible] [illegible] [illegible] [illegible] [illegible] [illegible] [illegible] [illegible] [illegible] [illegible] [illegible] [illegible]. They [illegible] [illegible] [illegible] [illegible] [illegible] [illegible] and they have [illegible].

2. The Ramification Problem

2.1 Indirect Effects of Actions

An invaluable advantage of action laws is that they allow to describe actions by their effects rather than by an exhaustive state transition table. However, the sole use of action laws quickly becomes unmanageable in complex domains, too. For action laws as they stand are supposed to be complete in that they specify the entire effect of an action. Yet although there are good reasons to assume that an action causes only a small number of *direct* changes, these in turn may initiate a long chain of *indirect* effects. Recall the action of toggling a switch, which, in the first place, causes nothing but a change of the switch's position. However, the switch may be part of an electric circuit so that, say, some light bulb is turned off as a side effect, which in turn may cause someone to hurt himself in a suddenly darkened room by running against a chair that, as a consequence, falls into a television set whose implosion activates the fire alarm and so on and so forth.[1]

Let us state the problem more precisely. Suppose we perform the action of toggling the switch in the simple electric circuit depicted in Fig. 2.1. In the first place, the action changes nothing but the position of the switch, which thus is the *direct* effect. Obviously, however, a now closed switch means that the light goes on.[2] This is an *indirect* effect of our action. This additional effect is a consequence of the general relation between the position of the switch and the state of the bulb, namely, light is on if and only if the switch is in the upper position. This relation is formally described by the fluent formula $\texttt{light} \equiv \texttt{up}(\texttt{s}_1)$. Fluent formulas of this kind, which are supposed

[1] A crucial question in this context concerns the distinction between indirect effects occurring during a single world's state transition step and those which deserve separate state transitions (also called "delayed" effects). E.g., the above may not only be described as "the fire alarm is activated in the successor state after having closed the switch," but also as, say, "the chair is falling in the successor state (and presumably hits the television set during the next state transition)." As a reasonable, albeit informal, guidance we suggest a single state transition should summarize what happens until some agent has the possibility to intervene by acting again (stopping the chair from falling, for instance). See also Section 2.7 for a discussion on indirect vs. delayed effects.

[2] We tacitly assume that battery and bulb are in working order; more to this in Chapter 4.

M. Thielscher: Challenges for Action Theories, LNAI 1775, pp. 17-83, 2000.

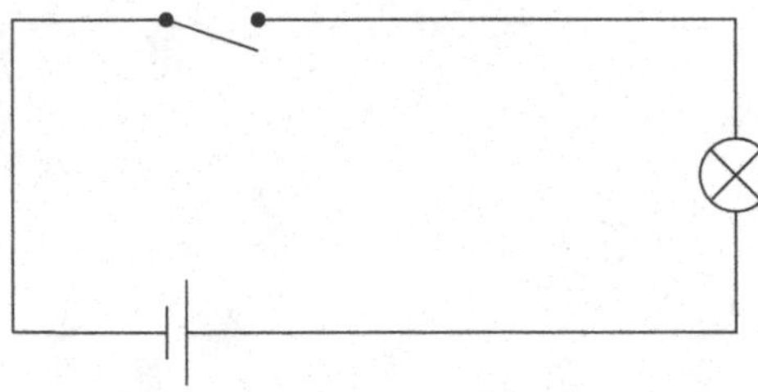

Figure 2.1. An electric circuit consisting of a battery, a switch, and a light bulb, which lights up if and only if the switch is closed. The two dynamic components are described by two fluents, both of which are false in the current state.

to be true in any state, are called *state constraints*. States satisfying all constraints of a domain shall be called *acceptable*.

State constraints, as we have seen, give rise to additional effects if violated after the direct effects of an action have been considered. Actions having indirect effects conflicts with the general assumption that nothing changes except what is mentioned in an action law. This can easily be seen with our example circuit: Let $S = \{\neg\texttt{up}(\texttt{s}_1), \neg\texttt{light}\}$ be the current state as depicted in Fig. 2.1. Performing a $\texttt{toggle}(\texttt{s}_1)$ action in S produces the state $S' = \{\texttt{up}(\texttt{s}_1), \neg\texttt{light}\}$ following the known action laws (1.2) and according to the definition of how to apply them. This is not the expected successor state, as is formally evident by its violating the underlying state constraint, $\texttt{light} \equiv \texttt{up}(\texttt{s}_1)$. Fortunately we were foresighted enough to call "preliminary" states like S'. The fact that often it does not suffice to compute, via the application of action laws, the mere direct effects is called the *Ramification Problem*.[3]

One straightforward 'solution' to the Ramification Problem is to circumvent it. That is, one could stick to the assumption that action laws be complete in specifying the entire effect. To this end, all indirect effects must somehow be compiled into the action laws. This procedure, however, bears two major problems demonstrating its inadequacy. First, it may require an enormous number of action laws to account for every possible combination of indirect effects. To see why, consider a model of an electric circuit where a distinguished switch is involved in n sub-circuits each of which additionally contains a switch-bulb pair; see Fig. 2.2. Defining all effects of toggling the separate switch solely by means of action laws then requires 2^{n+1} different laws, one for each possible combination of truth-values assigned to the

[3] The name shall suggest the picture that a state description literally ramifies into various directions, each following a possible chain of indirect effects.

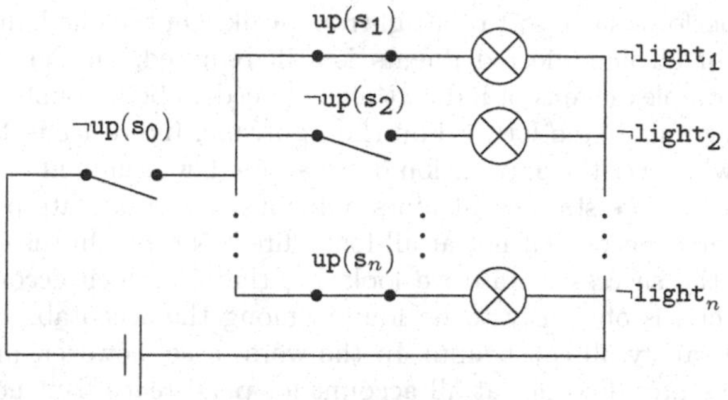

Figure 2.2. An electric circuit consisting of a battery, a separate switch $\mathtt{s_0}$, and n sub-circuits each containing a switch $\mathtt{s}_i$ and a light bulb $\mathtt{light}_i$. Exponentially many cases have to be considered when specifying the effect of closing $\mathtt{s_0}$ solely by means of action laws.

switch being operated and the other n switches. An arbitrary example is the following:

$$\begin{aligned}\mathtt{toggle(s_0)}\ \underline{\text{transforms}}\ &\{\neg\mathtt{up(s_0)}, \mathtt{up(s_1)}, \neg\mathtt{light_1}, \neg\mathtt{up(s_2)}, \neg\mathtt{light_2}, \ldots\} \\ \underline{\text{into}}\ &\{\mathtt{up(s_0)}, \mathtt{up(s_1)}, \mathtt{light_1}, \neg\mathtt{up(s_2)}, \neg\mathtt{light_2}, \ldots\}\end{aligned}$$

The other laws look no less unpleasant than this one. This is clearly far from what we consider an adequate specification. Full adequacy is achieved only if the original, simple specification (1.2) of action $\mathtt{toggle}(x)$ also applies to switch $\mathtt{s_0}$ in this more complex domain, and if the n state constraints $\mathtt{light}_i \equiv \mathtt{up(s_0)} \wedge \mathtt{up(s}_i\mathtt{)}$ do the rest. Notice that the size of this specification is linear (in n) as opposed to the exponential number of action laws.

The second problem with exhaustive action laws containing the entire effect is that the introduction of a new state constraint may require, in the worst case, a redefinition of the entire set of action laws used before. Just suppose yet another switch-bulb pair $\mathtt{up(s}_{n+1}\mathtt{)}, \mathtt{light}_{n+1}$ be added to the circuit of Fig. 2.2. Then all of the previously carefully designed action laws for $\mathtt{toggle(s_0)}$—and recall their exponential number—would need revision.

It so seems we have to think about real solutions to the Ramification Problem.

2.2 Minimizing Change

We have seen that the basic issue with the Ramification Problem is to resolve the conflict between the possibility of indirect effects and the postulate of persistence, i.e., that a fluent does not change unless it is explicitly so stated in the respective action law. The challenge, therefore, is to find a suitably weakened version of this postulate. The general difficulty with this is that one has to perform some sort of a tightrope walk. On the one hand, rigorous persistence of all non-affected fluents is still required; on the other hand, arbitrarily complex chains of indirect effects need to be accounted for.

Let us try a first step forward on the tightrope. Indirect effects are to be considered whenever the application of an action law results in a state which violates one or more state constraints. Obviously, such a state perfectly accounts for persistence—but not at all for indirect effects. On the other hand, notice that the successor state we look for, that is, which accounts for all indirect effects, is obviously to be found among the acceptable states, i.e., those which satisfy all constraints. In the worst case, however, an arbitrary one of these states does not at all account for persistence. But now suppose we combine the one extreme with the other. Namely, we select among all acceptable states the one that shares the most fluent literals with the preliminary successor state.[4] Then we have accounted both for indirect effects (since no constraint is violated) and for persistence to the largest possible extent (since no more fluents have changed than absolutely necessary to obtain an acceptable state). In other words, and from a more constructive perspective, the idea is to take the preliminary successor and to change the truth-values of as few as possible fluents—of course without touching one of the direct effects—until a state results that satisfies the state constraints.

To see how this approach works, recall the circuit of Fig. 2.1. We have seen that toggling the switch in the current state, viz. $\{\neg \texttt{up(s}_1\texttt{)}, \neg \texttt{light}\}$, yields the preliminary successor $S' = \{\texttt{up(s}_1\texttt{)}, \neg \texttt{light}\}$, which violates the underlying state constraint, $\texttt{light} \equiv \texttt{up(s}_1\texttt{)}$. The state closest to S' in satisfying this formula and in still containing the direct effect, $\texttt{up(s}_1\texttt{)}$, is $T = \{\texttt{up(s}_1\texttt{)}, \texttt{light}\}$. This is indeed the intended and intuitively expected successor state: The light went on as indirect effect of closing the switch.

The following formal account of this approach is based on a comparative notion of distance between states. We introduce the concept of *minimizing-change successors*, which are obtained according to the above description.

Definition 2.2.1. *Let $\mathcal{D}$ be a basic action domain. If S, T, T' are states, then T is* closer *to S than T', written $T \prec_S T'$, iff $T \setminus S \subsetneq T' \setminus S$. Let $\mathcal{C}$ be a set of state constraints, S a state which is acceptable (wrt. $\mathcal{C}$), and a an action. A state T is* minimizing-change successor *of S and a iff the following holds: There exists a preliminary successor S' of S and action a obtained through direct effect E such that*

1. *$E \subseteq T$,*
2. *T is acceptable, and*
3. *there is no $T' \prec_{S'} T$ such that $E \subseteq T'$ and T' is acceptable.* ■

[4] Actually this state 'closest' to the preliminary successor need not be unique, see below.

The condition $T \setminus S \subsetneq T' \setminus S$ for $T \prec_S T'$ states that S and T differ in strictly less fluent literals than S and T' do. The definition then says that a minimizing-change successor is a state which contains the direct effect of the action in question (clause 1), which satisfies the state constraints (clause 2), and which is closest to a preliminary successor S' in so doing (clause 3).

Example 2.2.1. Let $\mathcal{D}$ be the domain consisting of entity $\texttt{s}_1$, fluent names $\texttt{up}^1$ and $\texttt{light}^0$, action name $\texttt{toggle}^1$, and the familiar action laws

$$\texttt{toggle}(x)\ \underline{\text{transforms}}\ \{\texttt{up}(x)\}\ \underline{\text{into}}\ \{\neg\texttt{up}(x)\}$$
$$\texttt{toggle}(x)\ \underline{\text{transforms}}\ \{\neg\texttt{up}(x)\}\ \underline{\text{into}}\ \{\texttt{up}(x)\}$$

Along with state constraint $\mathcal{C} = \{\texttt{light} \equiv \texttt{up}(\texttt{s}_1)\}$, this formalizes our circuit of Fig. 2.1.

Let the current state be the acceptable $\{\neg\texttt{up}(\texttt{s}_1), \neg\texttt{light}\}$. The only preliminary successor state of S and $\texttt{toggle}(\texttt{s}_1)$ is $S' = \{\texttt{up}(\texttt{s}_1), \neg\texttt{light}\}$, obtained through direct effect $E = \{\texttt{up}(\texttt{s}_1)\}$. Now, there is just one acceptable state containing E, viz. $\{\texttt{up}(\texttt{s}_1), \texttt{light}\}$. Being the only candidate, this state is of course the one closest to S', hence it is the unique minimizing-change successor when performing $\texttt{toggle}(\texttt{s}_1)$ in the state $\{\neg\texttt{up}(\texttt{s}_1), \neg\texttt{light}\}$. ■

Our first approach to the Ramification Problem works fine with our small example domain. The interested reader may verify, for instance, that if in the successor state obtained above we toggle the switch again, then not only does it take its original position but also the light is off in the resulting minimizing-change successor. Or, suppose that in the current state depicted in Fig. 2.1 we perform a non-deterministic action to the effect that the switch may or may not get closed, then there are two minimizing-change successor states: Either the switch is up and light is on, or else the position of the switch has not changed and the light stays off.

For a general assessment of the approach of minimizing change, this observation is crucial: All fluents of a state constraint which are not among the direct effects have equal right to change their value in case this constraint is violated by the preliminary successor at hand. This appears to be no problem as long as only two fluents are involved in a constraint, one of which must have changed so that this constraint became violated.[5] As soon as a state constraint relates three or more fluents, however, it may be erroneous to consider equally possible all state adaptations that correct a violation. To illustrate this, we enhance our electric circuit by a second switch as shown in Fig. 2.3. This—at first glance innocent—modification has a surprising effect on the applicability of our first approach to the Ramification Problem.

Example 2.2.2. Let $\mathcal{D}$ be the domain consisting of entities $\texttt{s}_1$ and $\texttt{s}_2$, fluent names $\texttt{up}^1$ and $\texttt{light}^0$, action name $\texttt{toggle}^1$, and the following action laws.

[5] Surprisingly enough, even in this simple case the approach may lead to the wrong conclusion, e.g., if applied to a scenario discussed later, in Section 2.8.

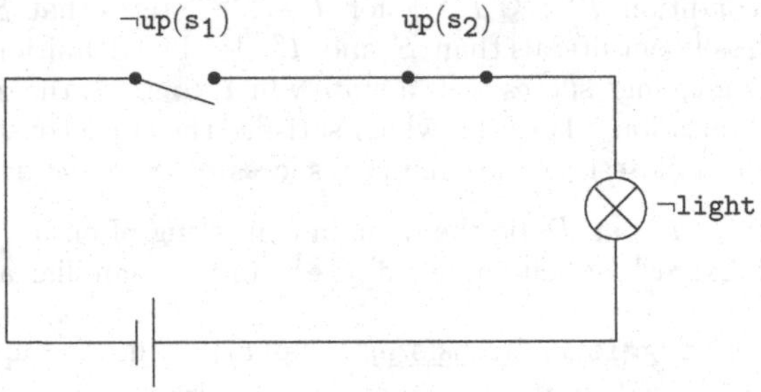

Figure 2.3. The electric circuit of Fig. 2.1 enhanced by a second switch. What would you expect as to the outcome of closing $\mathtt{s_1}$ in the current state?

$$\mathtt{toggle}(x) \ \underline{\text{transforms}} \ \{\mathtt{up}(x)\} \ \ \underline{\text{into}} \ \{\neg\mathtt{up}(x)\}$$
$$\mathtt{toggle}(x) \ \underline{\text{transforms}} \ \{\neg\mathtt{up}(x)\} \ \underline{\text{into}} \ \{\mathtt{up}(x)\}$$

The two switches and the light bulb being serially connected, we introduce the state constraint $\mathtt{light} \equiv \mathtt{up(s_1)} \wedge \mathtt{up(s_2)}$.

Let the current state be the acceptable $\{\neg\mathtt{up(s_1)}, \mathtt{up(s_2)}, \neg\mathtt{light}\}$ as depicted in Fig. 2.3. Now suppose $\mathtt{toggle(s_1)}$ be performed in this state, then the unique preliminary successor is $S' = \{\mathtt{up(s_1)}, \mathtt{up(s_2)}, \neg\mathtt{light}\}$, which is obtained through the direct effect $E = \{\mathtt{up(s_1)}\}$. This state violates the state constraint. There are two acceptable states containing E, namely, $T_1 = \{\mathtt{up(s_1)}, \mathtt{up(s_2)}, \mathtt{light}\}$ and $T_2 = \{\mathtt{up(s_1)}, \neg\mathtt{up(s_2)}, \neg\mathtt{light}\}$. As for the respective distance to S', we observe that $T_1 \setminus S' = \{\mathtt{light}\}$ and $T_2 \setminus S' = \{\neg\mathtt{up(s_2)}\}$. Thus neither $T_1 \prec_{S'} T_2$ nor $T_2 \prec_{S'} T_1$. Consequently, both T_1 and T_2 are minimizing-change successors. ■

So instead of coming to the obvious conclusion that the light must turn on as a side effect of toggling $\mathtt{s_1}$, an equally possible course of events is suggested, namely, that switch $\mathtt{s_2}$ leaves its position and the light stays off!

What is the reason for this unexpected outcome? A closer examination of the underlying state constraint reveals it. From $\mathtt{light} \equiv \mathtt{up(s_1)} \wedge \mathtt{up(s_2)}$ we can deduce that $\mathtt{up(s_1)} \supset \mathtt{light} \vee \neg\mathtt{up(s_2)}$ but not the stronger implication $\mathtt{up(s_1)} \supset \mathtt{light}$ (nor, of course, $\mathtt{up(s_1)} \supset \neg\mathtt{up(s_2)}$). In words, suppose we know that switch $\mathtt{s_1}$ is in the upper position, then it follows that light is on *or* switch $\mathtt{s_2}$ is down. It does not necessarily follow that the light is on. Therefore, if the state constraint becomes violated by $\mathtt{up(s_1)}$ becoming true, the mere disjunction $\mathtt{light} \vee \neg\mathtt{up(s_2)}$ needs to be satisfied. Obviously, this can be achieved either by changing $\neg\mathtt{light}$ to $\mathtt{light}$ or, *voilà*, by changing $\mathtt{up(s_2)}$ to $\neg\mathtt{up(s_2)}$. The pure state constraint does not allow to distinguish between these two possibilities in view of preferring the former, as one would like. Hence the two minimizing-change successor states.

There is a general principle behind the problem we have encountered with this example. Often a mere state constraint does not contain sufficient information as to what effects it is expected to trigger. From the perspective of logic, all fluent changes that correct a violation of a constraint are equal. Some of these changes, however, would never occur in reality as indirect effect. Consequently, additional domain knowledge is required that allows to distinguish the deductions which correspond to real effects from the mere logical consequences which have no equivalent in reality. This is a challenge of very general nature, which we will have to face at various places throughout the book. Our objective, therefore, is to develop a solution on the basis of a principle (yet to be found) which is as universal as the problem to be addressed.

2.3 Categorizing Fluents

At the end of the previous section we arrived at the insight that in general a mere state constraint is insufficient for a correct handling of indirect effects. This raises the issue of the nature of additionally required information. With regard to our example domain, it would of course be sufficient to add the very specific knowledge that "when closing the switch, an activation of the light bulb is likely to occur, as opposed to the other switch changing its position." Yet our aim is to find a general criterion that helps us telling apart the correct indirect effects. That is to say, we seek the *reason* for us expecting the light turns on rather than that the switch moves.

In fact, one essential difference between the light bulb and the switch is fairly apparent. Namely, the state of a switch can be affected only by direct operation. Its state is, in a certain sense, independent of the values of other fluents. In contrast, there are no means to directly operate the light bulb. Rather its state completely depends on the values of other fluents. From this perspective, it is no surprise that toggling a switch may indirectly affect the light but not the other switch. This observation suggests the following refinement of our first approach to the Ramification Problem.

In order that only expected indirect effects are generated, the fluents of a domain are divided into two categories. One of which consists of those fluents that represent state components which are manipulated by direct operation. The other category contains all fluents that represent state components which depend on other components. Let us call *primary* fluents of the former kind and *secondary* those of the latter. What fluent belongs to which category cannot, of course, be gathered from the state constraints. Therefore, the categorization needs to be given as part of the domain specification. On the basis of this additional information, whenever the procedure of minimizing change offers a choice between a primary and a secondary fluent, then the former is preferred to remain unchanged—hence the latter is preferred to adapt as indirect effect.

A formalization of this approach requires just a marginal refinement of our concept of minimizing-change successors (recall Definition 2.2.1). Namely, the notion of distance shall now respect the preference for minimizing change of primary fluents. Otherwise our definition remains unchanged.

Definition 2.3.1. *Let $\mathcal{D} = (\mathcal{E}, \mathcal{F}, \mathcal{A}, \mathcal{L})$ be a basic action domain. Furthermore, let $\mathcal{F}_p$ (*primary *fluents) and $\mathcal{F}_s$ (*secondary *fluents) be two sets of fluents such that $\mathcal{F}_p \cap \mathcal{F}_s = \{\}$ and $\mathcal{F}_p \cup \mathcal{F}_s$ is the set of all fluents composed of fluent names $\mathcal{F}$ and entities $\mathcal{E}$. If S, T, T' are states, then T is* closer *to S than T'* wrt. *$\mathcal{F}_p, \mathcal{F}_s$, written $T \prec_S T' |_{\mathcal{F}_p,\mathcal{F}_s}$, iff*

1. *$\|T \setminus S\| \cap \mathcal{F}_p \subsetneq \|T' \setminus S\| \cap \mathcal{F}_p$, or*
2. *$\|T \setminus S\| \cap \mathcal{F}_p = \|T' \setminus S\| \cap \mathcal{F}_p$ and $\|T \setminus S\| \cap \mathcal{F}_s \subsetneq \|T' \setminus S\| \cap \mathcal{F}_s$.*

Let C be a set of state constraints, S a state which is acceptable (wrt. C), and a an action. A state T is categorized minimizing-change successor *of S and a iff the following holds: There exists a preliminary successor S' of S and action a obtained through direct effect E such that*

1. *$E \subseteq T$,*
2. *T is acceptable, and*
3. *there is no $T' \prec_{S'} T |_{\mathcal{F}_p,\mathcal{F}_s}$ such that $E \subseteq T'$ and T' is acceptable.* ■

With the refined notion of closeness of states we first concentrate on primary fluents (clause 1) when comparing distances. Only in case the distances are equal with this regard, secondary fluents become the decisive factor (clause 2).

Example 2.3.1. Let domain $\mathcal{D}$ be as in Example 2.2.2 (recall the circuit of Fig. 2.3, with two switches). Suppose that $\mathcal{F}_p = \{\texttt{up}(x) : x \in \{\texttt{s}_1, \texttt{s}_2\}\}$ and $\mathcal{F}_s = \{\texttt{light}\}$ be the fluent categorization for $\mathcal{D}$. When performing $\texttt{toggle}(\texttt{s}_1)$ in the acceptable state $S = \{\neg\texttt{up}(\texttt{s}_1), \texttt{up}(\texttt{s}_2), \neg\texttt{light}\}$, the (unique) preliminary successor is $S' = \{\texttt{up}(\texttt{s}_1), \texttt{up}(\texttt{s}_2), \neg\texttt{light}\}$, which results through direct effect $E = \{\texttt{up}(\texttt{s}_1)\}$ as before. There exist two acceptable states containing E, namely, $T_1 = \{\texttt{up}(\texttt{s}_1), \texttt{up}(\texttt{s}_2), \texttt{light}\}$ and $T_2 = \{\texttt{up}(\texttt{s}_1), \neg\texttt{up}(\texttt{s}_2), \neg\texttt{light}\}$. From $\|T_1 \setminus S'\| \cap \mathcal{F}_p = \{\texttt{light}\} \cap \mathcal{F}_p = \{\}$ and from $\|T_2 \setminus S'\| \cap \mathcal{F}_p = \{\texttt{up}(\texttt{s}_2)\} \cap \mathcal{F}_p = \{\texttt{up}(\texttt{s}_2)\}$ we can conclude that $T_1 \prec_{S'} T_2 |_{\mathcal{F}_p,\mathcal{F}_s}$ (but not vice versa, of course). Hence T_1 is the unique categorized minimizing-change successor of $\{\neg\texttt{up}(\texttt{s}_1), \texttt{up}(\texttt{s}_2), \neg\texttt{light}\}$ and $\texttt{toggle}(\texttt{s}_1)$. ■

The distinction between primary and secondary fluents allows to recognize 'phantom' effects whenever these are obtained by changing a generally independent property where changing a dependent one would suffice to satisfy a state constraint. In this way we have avoided, for instance, the conclusion that a switch magically leaves its position where it was sufficient to turn on the light. On the other hand, it is obvious that this approach to the Ramification Problem relies on the existence of a suitable categorization for a domain. The following example with a newly extended electric circuit shows that this is not guaranteed.

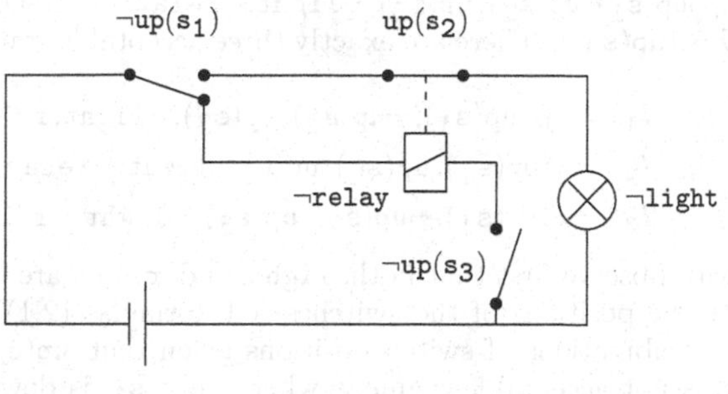

Figure 2.4. Yet another enhancement of our electric circuit: It now includes a third switch, $\mathtt{s_3}$, and a relay, whose state is represented by the nullary fluent $\mathtt{relay}$. When activated, the relay forces switch $\mathtt{s_2}$ downwards.

Example 2.3.2. Consider the electric circuit depicted in Fig. 2.4. Let this be modeled by domain $\mathcal{D}$ consisting of entities $\mathtt{s_1}$, $\mathtt{s_2}$, and $\mathtt{s_3}$; fluent names $\mathtt{up}^1$, $\mathtt{light}^0$, and $\mathtt{relay}^0$; action name $\mathtt{toggle}^1$; and the action laws

$$\mathtt{toggle}(x)\ \underline{\text{transforms}}\ \{\mathtt{up}(x)\}\ \underline{\text{into}}\ \{\neg\mathtt{up}(x)\}$$
$$\mathtt{toggle}(x)\ \underline{\text{transforms}}\ \{\neg\mathtt{up}(x)\}\ \underline{\text{into}}\ \{\mathtt{up}(x)\}$$

Let $\mathcal{C}$ be the three state constraints

$$\begin{aligned} \mathtt{light} &\equiv \mathtt{up}(\mathtt{s_1}) \wedge \mathtt{up}(\mathtt{s_2}) \\ \mathtt{relay} &\equiv \neg\mathtt{up}(\mathtt{s_1}) \wedge \mathtt{up}(\mathtt{s_3}) \\ \mathtt{relay} &\supset \neg\mathtt{up}(\mathtt{s_2}) \end{aligned} \tag{2.1}$$

The last formula states that the activated relay attracts switch $\mathtt{s_2}$. Let the current state be $S = \{\neg\mathtt{up}(\mathtt{s_1}), \mathtt{up}(\mathtt{s_2}), \neg\mathtt{up}(\mathtt{s_3}), \neg\mathtt{light}, \neg\mathtt{relay}\}$, as shown in Fig. 2.4.

Now, for a suitable account of possible indirect effects of toggling switches in this circuit we are supposed to categorize the involved fluents. This is straightforward in case of $\mathtt{up}(\mathtt{s_1})$ and $\mathtt{up}(\mathtt{s_3})$ (being primary) and $\mathtt{light}$ and $\mathtt{relay}$ (being secondary); it is easy to verify that any other choice would immediately cause undesired effects. But what about the second switch, represented by $\mathtt{up}(\mathtt{s_2})$? On the one hand, we naturally tend to consider it primary as before. For if $\mathtt{s_1}$ is toggled in the current state, then $\mathtt{s_2}$ is supposed not to change where light is expected to do so. On the other hand, there is reason to consider $\mathtt{up}(\mathtt{s_2})$ secondary. Namely, if we close switch $\mathtt{s_3}$ in the current state, then the relay gets activated and, hence, is expected to cause switch $\mathtt{s_2}$ leave its position. Formal examination reveals that this would not be the only categorized minimizing-change successor in this scenario had we $\mathcal{F}_p = \{\mathtt{up}(\mathtt{s_1}), \mathtt{up}(\mathtt{s_2}), \mathtt{up}(\mathtt{s_3})\}$ and $\mathcal{F}_s = \{\mathtt{light}, \mathtt{relay}\}$: The unique preliminary successor of state S from above and action $\mathtt{toggle}(\mathtt{s_3})$

is $S' = \{\neg\mathtt{up(s_1)}, \mathtt{up(s_2)}, \mathtt{up(s_3)}, \neg\mathtt{light}, \neg\mathtt{relay}\}$, obtained through direct effect $E = \{\mathtt{up(s_3)}\}$. There are exactly three acceptable states containing E, viz.

$$\begin{aligned} T_1 &= \{\neg\mathtt{up(s_1)}, \neg\mathtt{up(s_2)}, \mathtt{up(s_3)}, \neg\mathtt{light}, \mathtt{relay}\} \\ T_2 &= \{\mathtt{up(s_1)}, \mathtt{up(s_2)}, \mathtt{up(s_3)}, \mathtt{light}, \neg\mathtt{relay}\} \\ T_3 &= \{\mathtt{up(s_1)}, \neg\mathtt{up(s_2)}, \mathtt{up(s_3)}, \neg\mathtt{light}, \neg\mathtt{relay}\} \end{aligned}$$

To see why, observe first that both `light` and `relay` are completely determined by the positions of the switches (c.f. formulas (2.1)). There are four different combinations of switch positions given that $\mathtt{up(s_3)}$, one of which, however, is not acceptable, namely, where only $\mathtt{s_1}$ is down. As for the respective distance to S', we have

$$\begin{aligned} \|T_1 \setminus S'\| \cap \mathcal{F}_p &= \{\mathtt{up(s_2)}, \mathtt{relay}\} \cap \mathcal{F}_p = \{\mathtt{up(s_2)}\} \\ \|T_2 \setminus S'\| \cap \mathcal{F}_p &= \{\mathtt{up(s_1)}, \mathtt{light}\} \cap \mathcal{F}_p = \{\mathtt{up(s_1)}\} \\ \|T_3 \setminus S'\| \cap \mathcal{F}_p &= \{\mathtt{up(s_1)}, \mathtt{up(s_2)}\} \cap \mathcal{F}_p = \{\mathtt{up(s_1)}, \mathtt{up(s_2)}\} \end{aligned}$$

It follows that both $T_1 \prec_{S'} T_3|_{\mathcal{F}_p,\mathcal{F}_s}$ and $T_2 \prec_{S'} T_3|_{\mathcal{F}_p,\mathcal{F}_s}$ whereas T_1 and T_2 are not comparable wrt. S'. Hence the two are categorized minimizing-change successors of S and $\mathtt{toggle(s_3)}$. ■

The existence of the unintended successor, i.e., T_2, where switch $\mathtt{s_1}$ changes instead of switch $\mathtt{s_2}$, can be explained by the necessity of changing a primary fluent—aside from $\mathtt{up(s_3)}$, which was the direct effect—to arrive at an acceptable state. In the intended successor, i.e., T_1, this change concerns switch $\mathtt{s_2}$, triggered by the activation of the relay. This very activation is avoided, however, by moving switch $\mathtt{s_1}$ instead, as done in T_2. In both ways we respect the idea of minimizing change of primary fluents. Hence the two successor states.

The reason for this unexpected result is that we necessarily fail to assign a unique appropriate category to fluent $\mathtt{up(s_2)}$, whose role is twofold: On the one hand, it should be considered primary regarding the sub-circuit involving switch $\mathtt{s_1}$ and the light bulb, and on the other hand, it behaves like a secondary fluent as regards the relay. From a general perspective, this proves that it might be impossible to globally characterize a fluent either as always being 'active' or as always being 'passive.' The way a fluent behaves rather seems a more local property, depending on which of the related components one considers. In other words, a fluent can be active regarding one aspect and passive regarding another. As for our example, one might suggest just introducing an additional category of, say, *tertiary* fluents, $\mathcal{F}_t$, with even lower priority than secondary fluents. Then taking $\mathtt{up(s_1)}, \mathtt{up(s_3)} \in \mathcal{F}_p$, $\mathtt{up(s_2)}, \mathtt{relay} \in \mathcal{F}_s$, and $\mathtt{light} \in \mathcal{F}_t$ yields the expected unique resulting state, provided an appropriate extension of our notion of state distance wrt. a categorization (recall Definition 2.3.1). However, this particular classification requires a deeper analysis of possible direct and indirect effects in the electric circuit and seems, at first glance, quite unnatural. It is, moreover, not hard to imagine more complex domains demanding more and more categories, which heavily increases the difficulty of deciding which class a particular fluent should belong to. In particular, the addition of constraints to a domain specification may require a modification of the original categorization (as in our example, where up(s2) moved to $\mathcal{F}_s$ and light to $\mathcal{F}_t$)—which is to be considered inadequate.

2.4 Causal Relationships

The different approaches with which we attempted to tackle the Ramification Problem thus far mainly revealed that accounting for indirect effects of actions is more challenging an enterprise than it seems at first glance. Although the occurrence of indirect effects is triggered by the necessity to correct violations of state constraints, the very constraints turned out not to suffice for recognizing logically proposed effects that would never occur in reality. Providing additional domain knowledge in form of global classification of fluents proved too coarse in general as well. Whether or not a fluent is expected to change as indirect effect is a more local property, depending on the context. What we need, therefore, is a concept for generating indirect effects on the basis of knowledge whose structure respects and reflects this context-sensitivity. The notion of *causal relationships* will serve this purpose. Being a key concept of this book, these relationships will be introduced and thoroughly analyzed in this and the following four sections.

2.4.1 Causal Relationships and Their Application

Causal relationships are formalizations of the circumstances under which the occurrence of a single indirect effect is to be expected. Two components constitute these circumstances. One of which describes the context required for the indirect effect. The context is represented by a fluent formula that needs to be satisfied in a state in order that the causal relationship applies. The second component is a particular single effect whose prior occurrence *causes* (hence the name) the indirect effect in question. This triggering effect itself may have been previously obtained as indirect effect, if it was not among the direct effects of the action.

The distinction between a context and a particular triggering effect is essential. The reason for this will be elucidated in a moment. First, let us consider an example of a causal relationship, which occurs in the electric circuit with just two switches and the light bulb of Fig. 2.3. Suppose switch s1 gets closed as a direct effect of some action. This effect is expected to cause the light go on as indirect effect, provided switch s2 is closed, too. We write this causal relationship as

$$\texttt{up(s}_1\texttt{)}\ \underline{\text{causes}}\ \texttt{light}\ \underline{\text{if}}\ \texttt{up(s}_2\texttt{)} \tag{2.2}$$

Here $\mathtt{up(s_1)}$ is the effect whose appearance triggers indirect effect $\mathtt{light}$ in the context defined by the atomic fluent formula $\mathtt{up(s_2)}$. Of course, the analogue is true as well in this particular domain, viz.

$$\mathtt{up(s_2)}\ \underline{\text{causes}}\ \mathtt{light}\ \underline{\text{if}}\ \mathtt{up(s_1)} \tag{2.3}$$

Yet this symmetry is not always true, as we will see later. In any case, the two causal relationships express different things: The former applies whenever $\mathtt{up(s_1)}$ became true with $\mathtt{up(s_2)}$ already being true, as opposed to the latter, which applies whenever $\mathtt{up(s_2)}$ became true while $\mathtt{up(s_1)}$ holds.[6]

Causal relationships shall be employed to generate additional, indirect effects of actions after having obtained the direct effects through the application of an action law. Formally, causal relationships operate on state-effect pairs (S, E) where S is some current 'intermediate' state, a preliminary successor, for instance, and E contains all direct and indirect effects that have been generated so far. As an example, recall that $S' = \{\mathtt{up(s_1)}, \mathtt{up(s_2)}, \neg\mathtt{light}\}$ is the preliminary successor of toggling the first switch in the state depicted in Fig. 2.3. State S' is obtained through direct effect $E = \{\mathtt{up(s_1)}\}$. The causal relationship (2.2) above is applicable to (S', E) on account of both $\mathtt{up(s_1)}$ being part of effect E and $\mathtt{up(s_2)}$, the context, being true in S'. This relationship implies that S' should be modified so as to account for the indirect effect $\mathtt{light}$, which yields the state $S'' = \{\mathtt{up(s_1)}, \mathtt{up(s_2)}, \mathtt{light}\}$. In addition, we augment E by our new effect, $\mathtt{light}$. Altogether the result of applying the causal relationship to $(S', \{\mathtt{up(s_1)}\})$ is the state-effect pair $(S'', \{\mathtt{up(s_1)}, \mathtt{light}\})$. Notice that our second causal relationship, (2.3), should not be applicable to (S', E) despite $\mathtt{up(s_1)}$ being true in S', because effect $\mathtt{up(s_2)}$ is not contained in E.

The reason for maintaining the second component, E, is that identical intermediate states (such as S') can often be reached by different effects, each of which may require diverse, sometimes opposite treatment, as the following example illustrates.

Example 2.4.1. Suppose two switches $\mathtt{s_1}, \mathtt{s_2}$ be tightly coupled by a spring so that they are always in the same position; see Fig. 2.5. This is reflected by the state constraint $\mathtt{up(s_1)} \equiv \mathtt{up(s_2)}$. As a consequence, closing or opening either switch has the indirect effect that the other switch closes or opens, respectively, as well to release the tension of the spring. This is expressed by these four causal relationships:

$$\begin{array}{rcrcl} \mathtt{up(s_1)} & \underline{\text{causes}} & \mathtt{up(s_2)} & \underline{\text{if}} & \top \\ \mathtt{up(s_2)} & \underline{\text{causes}} & \mathtt{up(s_1)} & \underline{\text{if}} & \top \\ \neg\mathtt{up(s_1)} & \underline{\text{causes}} & \neg\mathtt{up(s_2)} & \underline{\text{if}} & \top \\ \neg\mathtt{up(s_2)} & \underline{\text{causes}} & \neg\mathtt{up(s_1)} & \underline{\text{if}} & \top \end{array} \tag{2.4}$$

[6] If both switches get closed as direct or indirect effects, then of course either relationship is applicabl and the result is the same, namely, light.

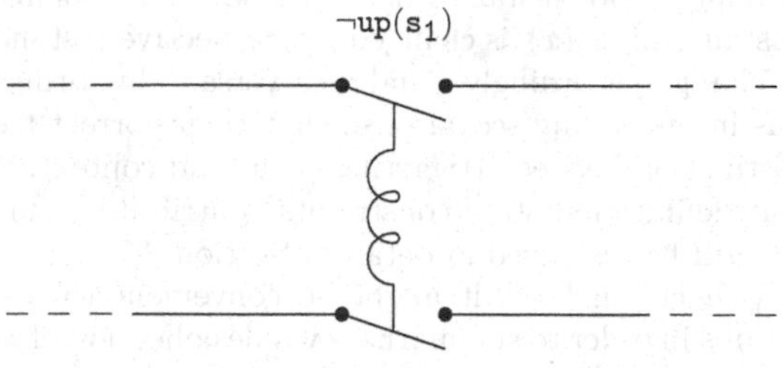

Figure 2.5. Two switches are mechanically connected through a tight spring, whose tension does not allow the switches occupy different positions.

Now, suppose we switch up the first switch in the state $\{\neg\mathtt{up}(\mathtt{s}_1), \neg\mathtt{up}(\mathtt{s}_2)\}$. This yields the preliminary successor $S' = \{\mathtt{up}(\mathtt{s}_1), \neg\mathtt{up}(\mathtt{s}_2)\}$. On the other hand, suppose we switch down the second switch in the state $\{\mathtt{up}(\mathtt{s}_1), \mathtt{up}(\mathtt{s}_2)\}$. This obviously yields the very same preliminary successor S'. Nonetheless the expected outcomes in these two situations differ considerably: In the first case the final result should be that both $\mathtt{s}_1$ and $\mathtt{s}_2$ are in the upper position, as opposed to the second case where both are expected down! This distinction can only be made by referring to the differing direct effects, viz. $E_1 = \{\mathtt{up}(\mathtt{s}_1)\}$ as opposed to $E_2 = \{\neg\mathtt{up}(\mathtt{s}_2)\}$. The former enables only the application of the very first relationship in (2.4) to the intermediate result, the latter only the application of the very last one. The two resulting state-effect pairs are

$$(\{\mathtt{up}(\mathtt{s}_1), \mathtt{up}(\mathtt{s}_2)\}, \{\mathtt{up}(\mathtt{s}_1), \mathtt{up}(\mathtt{s}_2)\})$$
$$(\{\neg\mathtt{up}(\mathtt{s}_1), \neg\mathtt{up}(\mathtt{s}_2)\}, \{\neg\mathtt{up}(\mathtt{s}_2), \neg\mathtt{up}(\mathtt{s}_1)\})$$

Thus in either case we obtain the intended successor state. ■

Incidentally, this example also illustrates the necessity of distinguishing triggering effects from contexts in causal relationships: There is an obviously crucial difference between the situation where $\mathtt{up}(\mathtt{s}_1)$ became true in the context $\neg\mathtt{up}(\mathtt{s}_2)$ and the situation where $\neg\mathtt{up}(\mathtt{s}_2)$ became true in the context $\mathtt{up}(\mathtt{s}_1)$.[7]

Curious as the last example may seem at first glance, such tight coupling between fluents frequently occurs in an important class of state constraints. So-called "definitional" constraints define, for the sake of convenience, a fluent as abbreviation for a complex fluent formula. The constraint

[7] Let us add the marginal note that dividing the condition for the occurrence of an indirect effect into two components matches the distinction often made in philosophical accounts of causality between so-called "triggering" and "predisposal" causes.

$\forall x\,[\,\mathtt{down}(x) \equiv \neg\mathtt{up}(x)\,]$, for instance, defines down as not being up. Clearly, whenever an instance of $\mathtt{up}(x)$ is changed the respective instance of $\mathtt{down}(x)$ is supposed to change accordingly—and vice versa. This brings about similar situations as in our spring scenario, so that their correct treatment, too, relies on the distinction between triggering effect and context. In passing, let us mention that definitional state constraints contribute to another special problem, which will be discussed in detail in Section 2.7.

Much like with action laws, it might be convenient to use variables in causal relationships in order to summarize a whole collection of similar ground instances. E.g., suppose all switches of an electric circuit and a light bulb are serially connected, then

$$\mathtt{up}(x)\ \underline{\text{causes}}\ \mathtt{light}\ \underline{\text{if}}\ \forall y.\,\mathtt{up}(y)$$

states that having switched up any switch x causes light provided all switches are (now) in the upper position. With this we arrive at the following formal definition of causal relationships and their application to state-effect pairs.

Definition 2.4.1. *Let $\mathcal{E}$ and $\mathcal{F}$ be sets of entities and fluent names, respectively. A* causal relationship *is of the form*

$$\varepsilon\ \underline{\text{causes}}\ \varrho\ \underline{\text{if}}\ \Phi$$

where Φ (the context*) is a fluent formula and both ε (the* triggering effect*) and ϱ (the* ramification*) are fluent expressions.*[8]

Let (S, E) be a pair consisting of a state S and a set of fluent literals E. Furthermore, let $r = \varepsilon\ \underline{\text{causes}}\ \varrho\ \underline{\text{if}}\ \Phi$ be a causal relationship, and let $\overline{x}$ denote a sequence of all free variables occurring in ε, ϱ, or Φ. Then a ground instance $r[\overline{e}]$ is applicable *to (S, E) iff $\varepsilon[\overline{e}] \in E$ and $\Phi[\overline{e}] \wedge \neg\varrho[\overline{e}]$ is true in S. The* application *of $r[\overline{e}]$ to (S, E) yields the pair (S', E') where $S' = (S \setminus \{\neg\varrho[\overline{e}]\}) \cup \{\varrho[\overline{e}]\}$ and $E' = (E \setminus \{\neg\varrho[\overline{e}]\}) \cup \{\varrho[\overline{e}]\}$. If $\mathcal{R}$ is a set of causal relationships, then by $(S, E) \leadsto_{\mathcal{R}} (S', E')$ we denote the existence of a relationship in $\mathcal{R}$ whose application to (S, E) yields (S', E').* ■

To summarize, a causal relationship $\varepsilon\ \underline{\text{causes}}\ \varrho\ \underline{\text{if}}\ \Phi$ is applicable if context Φ holds, the indirect effect-to-be ϱ is currently false, and if its cause ε is among the current effects. A relationship is applied by changing $\neg\varrho$ to ϱ in the current state and adding ϱ to the current effects. In order that the latter does not produce an inconsistency among the effects, a possibly preceding effect $\neg\varrho$ is withdrawn.[9] This precaution guarantees that whenever $(S, E) \leadsto_{\mathcal{R}} (S', E')$ with S being a state and E being consistent, then S' is a state and E' is consistent, too.

[8] Recall that fluent expressions are of the form $(\neg)f(t_1, \ldots, t_n)$ with f being a fluent name and t_i a variable or entity.

[9] It may sound strange to presuppose the occurrence of such situations instead of disregarding them, or even excluding them for that matter. Yet later, in the following but one section, we will meet an example scenario where this withdrawing a previously obtained indirect effect is necessary and perfectly reasonable.

Thus far we have seen how applying a single causal relationship generates one particular indirect effect. The direct effects of an action may of course give rise to several indirect effects. Moreover, these effects may in turn cause further effects, possibly triggering even more and so on and so forth. Closing switch $\mathtt{s_3}$ in our circuit involving the relay (Fig. 2.4), for example, causes an activation of that relay, which in turn triggers an indirect effect by attracting switch $\mathtt{s_2}$. These chains of indirect effects are modeled by serial application of causal relationships.

Example 2.4.2. Let $\mathcal{D}$ be the basic action domain formalizing the electric circuit of Fig. 2.4 as in Example 2.3.2. As for the various causal dependencies among the components, we notice, first, that light turns on if either of the two switches $\mathtt{s_1}$ and $\mathtt{s_2}$ is caused up with the other one already being in its upper position. Conversely, light is off if either of these switches is caused down, regardless of the other switch's position. This is reflected by introducing the following four causal relationships:

$$\begin{array}{ll} \mathtt{up(s_1)}\ \underline{\text{causes}}\ \mathtt{light}\ \underline{\text{if}}\ \mathtt{up(s_2)} & \neg\mathtt{up(s_1)}\ \underline{\text{causes}}\ \neg\mathtt{light}\ \underline{\text{if}}\ \top \\ \mathtt{up(s_2)}\ \underline{\text{causes}}\ \mathtt{light}\ \underline{\text{if}}\ \mathtt{up(s_1)} & \neg\mathtt{up(s_2)}\ \underline{\text{causes}}\ \neg\mathtt{light}\ \underline{\text{if}}\ \top \end{array} \tag{2.5}$$

Analogously, the state of the relay causally depends upon the two switches $\mathtt{s_1}$ and $\mathtt{s_3}$ as follows:

$$\begin{array}{ll} \neg\mathtt{up(s_1)}\ \underline{\text{causes}}\ \mathtt{relay}\ \underline{\text{if}}\ \mathtt{up(s_3)} & \mathtt{up(s_1)}\ \underline{\text{causes}}\ \neg\mathtt{relay}\ \underline{\text{if}}\ \top \\ \mathtt{up(s_3)}\ \underline{\text{causes}}\ \mathtt{relay}\ \underline{\text{if}}\ \neg\mathtt{up(s_1)} & \neg\mathtt{up(s_3)}\ \underline{\text{causes}}\ \neg\mathtt{relay}\ \underline{\text{if}}\ \top \end{array} \tag{2.6}$$

Finally, the activated relay forces switch $\mathtt{s_2}$ be down, that is,

$$\mathtt{relay}\ \underline{\text{causes}}\ \neg\mathtt{up(s_2)}\ \underline{\text{if}}\ \top \tag{2.7}$$

Now, let $S = \{\neg\mathtt{up(s_1)}, \mathtt{up(s_2)}, \neg\mathtt{up(s_3)}, \neg\mathtt{light}, \neg\mathtt{relay}\}$ be the current state, as depicted in Fig. 2.4. Performing the action $\mathtt{toggle(s_3)}$ yields the following state-effect pair.

$$(\{\neg\mathtt{up(s_1)}, \mathtt{up(s_2)}, \mathtt{up(s_3)}, \neg\mathtt{light}, \neg\mathtt{relay}\}, \{\mathtt{up(s_3)}\})$$

The one and only applicable causal relationship is the bottom left one in (2.6), which activates the relay:

$$(\{\neg\mathtt{up(s_1)}, \mathtt{up(s_2)}, \mathtt{up(s_3)}, \neg\mathtt{light}, \mathtt{relay}\}, \{\mathtt{up(s_3)}, \mathtt{relay}\})$$

As a consequence of this indirect effect, the relationship of (2.7) is now applicable, which results in

$$(\{\neg\mathtt{up(s_1)}, \neg\mathtt{up(s_2)}, \mathtt{up(s_3)}, \neg\mathtt{light}, \mathtt{relay}\}, \{\mathtt{up(s_3)}, \mathtt{relay}, \neg\mathtt{up(s_2)}\})$$

This state-effect pair allows no further application of causal relationships. Incidentally, its first component is acceptable wrt. the underlying state constraints (c.f. formulas (2.1)) and constitutes the (unique) resulting state expected when we close the third switch in initial state S. ■

By now it should have become clear how causal relationships are used to address the Ramification Problem. Starting with some preliminary successor state, we (non-deterministically) select and (serially) apply these relationships, each of which accommodates a single indirect effect. Provided the underlying set of causal relationships is designed so as to suitably represent real causal dependencies in a domain, all effects thus obtained are reasonable from the standpoint of causality. Whenever the application of a series finally produces an acceptable state, then the latter is taken as possible overall successor. Notice that each fluent which holds in the initial state and which is not touched at any step of this procedure remains unchanged. This approach therefore accounts both for rigorous persistence of non-affected parts on the one hand and for arbitrarily complex chains of indirect effects on the other hand.

In what follows, we say that a sequence of causal relationships $r_1, \ldots, r_n$ $(n \geq 0)$ is applicable to a pair (S_0, E_0) iff we can find n pairs $(S_1, E_1), \ldots,$ (S_n, E_n) such that for each $1 \leq i \leq n$, r_i is applicable to (S_{i-1}, E_{i-1}) yielding (S_i, E_i). We adopt a standard notation in writing $(S, E) \overset{*}{\leadsto}_{\mathcal{R}} (S_n, E_n)$ to indicate the existence of a (finite, possibly empty) sequence of causal relationships in $\mathcal{R}$ which is applicable to (S, E) with final result (S_n, E_n). The following is a formal definition of the notion of successor states on the basis of causal relationships.

Definition 2.4.2. *Let $\mathcal{E}$, $\mathcal{F}$, $\mathcal{A}$, and $\mathcal{L}$ be sets of entities, fluent names, action names, and action laws, respectively. Furthermore, let $\mathcal{C}$ be a set of state constraints and $\mathcal{R}$ a set of causal relationships. If S is an acceptable state and a an action, then a state S' is a* causal successor *of S and a iff the following holds: Set $\mathcal{L}$ contains an applicable instance a <u>transforms</u> C <u>into</u> E of an action law such that*

1. *$((S \setminus C) \cup E, E) \overset{*}{\leadsto}_{\mathcal{R}} (S', E')$ for some E', and*
2. *S' is acceptable.* ■

For a subtle reason to be revealed in Section 2.7 we are not yet completely satisfied with this definition as a general solution to the Ramification Problem. This is why we postpone the adaptation of the notions of interpretations and models to this later section.

The way causal relationships are employed makes no presuppositions as to the order in which they are applied. The two relationships used in our relay example to account for the indirect effects of closing the third switch can, however, be applied in only one order. For the chronologically second requires the result of the first, viz. relay activation, as the triggering effect. On the other hand, in general there may exist different ways of sequentially applying causal relationships. The formalization of our example circuit with n parallel sub-circuits each consisting of a switch-bulb pair (recall Fig. 2.2) shall illustrate this.

Example 2.4.3. Let $\mathcal{D}_n$ $(n \geq 2)$ be the basic action domain consisting of entities $\mathtt{s}_0, \mathtt{s}_1, \ldots, \mathtt{s}_n$, fluent names $\mathtt{up}^1, \mathtt{light_1}^0, \ldots, \mathtt{light}_n{}^0$, action name

$\mathtt{toggle}^1$, and action laws

$$\mathtt{toggle}(x)\ \underline{\text{transforms}}\ \{\mathtt{up}(x)\}\ \underline{\text{into}}\ \{\neg\mathtt{up}(x)\}$$
$$\mathtt{toggle}(x)\ \underline{\text{transforms}}\ \{\neg\mathtt{up}(x)\}\ \underline{\text{into}}\ \{\mathtt{up}(x)\}$$

Suppose further given the n state constraints $\mathtt{light}_i \equiv \mathtt{up}(\mathtt{s_0}) \wedge \mathtt{up}(\mathtt{s}_i)$ $(1 \leq i \leq n)$ and, for each $i = 1, \ldots, n$, these four causal relationships:

$$\begin{array}{ll} \mathtt{up}(\mathtt{s_0})\ \underline{\text{causes}}\ \mathtt{light}_i\ \underline{\text{if}}\ \mathtt{up}(\mathtt{s}_i) & \neg\mathtt{up}(\mathtt{s_0})\ \underline{\text{causes}}\ \neg\mathtt{light}_i\ \underline{\text{if}}\ \top \\ \mathtt{up}(\mathtt{s}_i)\ \underline{\text{causes}}\ \mathtt{light}_i\ \underline{\text{if}}\ \mathtt{up}(\mathtt{s_0}) & \neg\mathtt{up}(\mathtt{s}_i)\ \underline{\text{causes}}\ \neg\mathtt{light}_i\ \underline{\text{if}}\ \top \end{array} \tag{2.8}$$

(Notice the size of this specification, which is linear in n as opposed to the exponential number of action laws needed when formalizing this domain without the concept of indirect effects.) Now, suppose all switches are closed except for $\mathtt{s_0}$ and, hence, all light bulbs are off. That is, let $S = \{\neg\mathtt{up}(\mathtt{s_0}), \mathtt{up}(\mathtt{s_1}), \ldots, \mathtt{up}(\mathtt{s}_n), \neg\mathtt{light_1}, \ldots, \neg\mathtt{light}_n\}$ be the current, acceptable state. Performing action $\mathtt{toggle}(\mathtt{s_0})$ in this state yields the preliminary successor $S' = \{\mathtt{up}(\mathtt{s_0}), \mathtt{up}(\mathtt{s_1}), \ldots, \mathtt{up}(\mathtt{s}_n), \neg\mathtt{light_1}, \ldots, \neg\mathtt{light}_n\}$ along with the direct effect $E = \{\mathtt{up}(\mathtt{s_0})\}$. Apparently, the state-effect pair (S', E) allows the application of all n causal relationships of the form $\mathtt{up}(\mathtt{s_0})\ \underline{\text{causes}}\ \mathtt{light}_i\ \underline{\text{if}}\ \mathtt{up}(\mathtt{s}_i)$ where $i = 1, \ldots, n$. Whichever of these is executed first, the other $n-1$ relationships remain applicable in the resulting state-effect pair, and so forth. Thus there exist $n!$ different sequences, all of which obviously result in the same causal successor state, namely, $\{\mathtt{up}(\mathtt{s_0}), \mathtt{up}(\mathtt{s_1}), \ldots, \mathtt{up}(\mathtt{s}_n), \mathtt{light_1}, \ldots, \mathtt{light}_n\}$, i.e., where all light bulbs are now active. ∎

All application sequences coming to the identical result in this example raises the question whether order irrelevance holds in general. The following proposition states that indeed any permutation of a sequence of causal relationships arrives at the same conclusion, provided the permuted sequence is applicable.

Proposition 2.4.1. *Let $\mathcal{E}$ and $\mathcal{F}$ be sets of entities and fluent names, respectively, S_0 be a state, and E_0 a set of fluent literals. Furthermore, let $r_1, \ldots, r_n$ be a sequence of causal relationships ($n \geq 0$) which is applicable to (S_0, E_0) and which yields*

$$(S_0, E_0) \leadsto_{\{r_1\}} (S_1, E_1) \leadsto_{\{r_2\}} \ldots \leadsto_{\{r_n\}} (S_n, E_n)$$

Then, for any permutation $r_{\pi(1)}, \ldots, r_{\pi(n)}$ which is also applicable to (S_0, E_0) and which yields

$$(S_0, E_0) \leadsto_{\{r_{\pi(1)}\}} (S'_1, E'_1) \leadsto_{\{r_{\pi(2)}\}} \ldots \leadsto_{r_{\{\pi(n)\}}} (S'_n, E'_n)$$

we have $S_n = S'_n$ and $E_n = E'_n$.

Proof. Let f be an arbitrary but fixed fluent. By k_f^+ we denote the number (possibly zero) of relationships $r_i = \varepsilon_i$ causes f if Φ_i, and by k_f^- the number (possibly zero, too) of relationships $r_j = \varepsilon_j$ causes $\neg f$ if Φ_j $(1 \leq i, j \leq n)$. Since a causal relationship ε_i causes f if Φ_i can only be applied to some (S_{i-1}, E_{i-1}) if $\neg f \in S_{i-1}$ (and vice versa in case of indirect effect $\neg f$), the values for k_f^+ and k_f^- determine the final truth-value of f as follows:

1. If $f \in S$, then either $k_f^+ = k_f^-$ or $k_f^+ = k_f^- - 1$. In the former case we have $f \in S_n$. We also have $f \in E_n$ if $k_f^+ > 0$, otherwise $f \notin E_n$ and $\neg f \notin E_n$ since no causal relationship affects f. In the latter case we have both $\neg f \in S_n$ and $\neg f \in E_n$.
2. If $\neg f \in S$, then either $k_f^+ = k_f^-$ or $k_f^+ = k_f^- + 1$. In the former case we have $\neg f \in S_n$. We also have $\neg f \in E_n$ if $k_f^- > 0$, otherwise $\neg f \notin E_n$ and $f \notin E_n$ since no causal relationship affects f. In the latter case we have both $f \in S_n$ and $f \in E_n$.

Since the permutation $r_{\pi(1)}, \ldots, r_{\pi(n)}$ contains exactly the same causal relationships as the original sequence, they do not differ in the values for k_f^+ and k_f^-. Thus S_n and S'_n agree as far as f is concerned, and so do E_n and E'_n. Fluent f being an arbitrary choice proves the claim. *Qed.*

While this result proves general invariance with regard to applicable permutations of causal relationship sequences, it does *not* imply confluence of the whole procedure in general. In fact, a different selection at the beginning may allow for the application of a completely different collection of causal relationships. If two chains of relationships do not contain identical elements, then neither is a permutation of the other and our Proposition 2.4.1 does not apply. This is, however, not at all a drawback or even fallacy, as one might guess at first. Rather it allows to accommodate actions which are deterministic as far as their direct effects are concerned but non-deterministic as regards the indirect effects they possibly trigger. That is, even if a unique preliminary successor exists, performing an action in some state may admit more than one possible causal successor state. An example for an action that is non-deterministic in this sense occurs in a domain to be presented in Section 2.6.

The opposite to the existence of multiple successor states is that no successor at all can be found despite one or more action laws are applicable to the state at hand. That is, no chain of causal relationships manages to transform a preliminary successor into an acceptable state. This hints at additional, implicit preconditions for the action in question—preconditions which derive from state constraints. Section 2.8 will be devoted to this phenomenon. In the following section, we first raise another central issue, namely, how causal relationships and the underlying state constraints are related—in particular, we seek a way to automatically extract the former from the latter.

2.5 Influence Information

The causality-based approach to the Ramification Problem relies, to state the obvious, on the underlying set of causal relationships being suitable for the domain at hand. This set should be sound in that each element represents an intuitively plausible causal relation, and it should also be complete in covering all conceivable indirect effects that derive from the given state constraints. The various causal relationships we used in our relay example (c.f. (2.5)–(2.7)), for instance, arguably constitute a suitable collection for this domain. Apparently there is a tight correspondence between these nine relationships and the three underlying state constraints (c.f. formulas (2.1)). For example, causal relationship $\mathtt{up(s_3)}$ causes $\mathtt{relay}$ if $\neg\mathtt{up(s_1)}$ derives from state constraint $\mathtt{relay} \equiv \neg\mathtt{up(s_1)} \wedge \mathtt{up(s_3)}$, which entails the implication $\neg\mathtt{up(s_1)} \wedge \mathtt{up(s_3)} \supset \mathtt{relay}$. Generally we have that for any ε causes ϱ if Φ of our nine relationships, the corresponding fluent formula $\Phi \wedge \varepsilon \supset \varrho$ can be deduced from one of the state constraints.

This observation suggests that causal relationships might be somehow automatically extractible from a set of state constraints. On the other hand, most of our discussion prior to introducing causal relationships centered around the problem that state constraints contain insufficient information to decide what are the indirect effects that can possibly occur in reality. To restate the problem, notice that our example constraint $\mathtt{relay} \equiv \neg\mathtt{up(s_1)} \wedge \mathtt{up(s_3)}$ also entails implication $\neg\mathtt{relay} \wedge \mathtt{up(s_3)} \supset \mathtt{up(s_1)}$. The corresponding causal relationship $\mathtt{up(s_3)}$ causes $\mathtt{up(s_1)}$ if $\neg\mathtt{relay}$, however, does not hold. There is an information gap between the mere state constraints and the valid causal relationships they determine. In the present section, we concern ourselves with filling this gap without the necessity of drawing up, all by hand, the set of causal relationships. That is, we seek a general method that allows a more compact providing of the additionally required information.

In the previous but one section we discussed the range of applicability of tackling the Ramification Problem by categorizing fluents as either primary or secondary. Our analysis showed that even in simple domains it can be impossible to globally characterize a fluent either as one which changes independently of all other fluents or as one which is generally manipulable by others. It turned out to be a local property whether or not a fluent is expected to change as indirect effect. In our example circuit involving the relay, for instance, the second switch is expected independent of switch $\mathtt{s_1}$ but may certainly be affected by the relay.[10] In other words, fluent $\mathtt{up(s_1)}$ has no direct influence on fluent $\mathtt{up(s_2)}$, as opposed to fluent $\mathtt{relay}$. Domain knowledge stating whether a particular fluent may or may not directly affect another particular fluent should therefore help telling apart the correct causal

[10] Actually it is more precise to say the second switch does not *directly* depend on switch $\mathtt{s_1}$. Indirectly it does so, for the latter controls the relay, which in turn may control the second switch.

relationships. Formally, this so-called *influence information* is specified as a binary relation on the set of fluents.

Definition 2.5.1. *Let $\mathcal{E}$ and $\mathcal{F}$ be sets of entities and fluent names, respectively. An irreflexive binary relation $\mathcal{I}$ on the set of fluents is called* influence information. ■

If $(f_1, f_2) \in \mathcal{I}$, then this is intended to denote that a change of fluent f_1's truth-value potentially affects the truth-value of fluent f_2. For convenience, a pair in $\mathcal{I}$ may contain variables, thus representing all of its ground instances.

Example 2.5.1. Consider the entities $\mathcal{E} = \{\texttt{s}_1, \texttt{s}_2, \texttt{s}_3\}$ along with the fluents $\mathcal{F} = \{\texttt{up}^1, \texttt{light}^0, \texttt{relay}^0\}$. The two switches $\texttt{s}_1$ and $\texttt{s}_2$ may influence the light but not vice versa nor do they mutually interfere (c.f. Fig. 2.4). Likewise, $\texttt{s}_1$ and $\texttt{s}_3$ may influence the relay, which in turn potentially affects the position of $\texttt{s}_2$. Thus the correct influence information $\mathcal{I}$ is

$$\begin{array}{c}\{(\texttt{up}(\texttt{s}_1), \texttt{light}), (\texttt{up}(\texttt{s}_2), \texttt{light}), (\texttt{up}(\texttt{s}_1), \texttt{relay}), (\texttt{up}(\texttt{s}_3), \texttt{relay}), \\ (\texttt{relay}, \texttt{up}(\texttt{s}_2))\}\end{array} \quad (2.9)$$

Notice how the twofold nature of $\texttt{up}(\texttt{s}_2)$ is reflected in this influence information: The fluent occurs both as first (the active) component in the pair $(\texttt{up}(\texttt{s}_2, \texttt{light})$ and as second (the passive) component in the pair $(\texttt{relay}, \texttt{up}(\texttt{s}_2))$. This is why it was impossible to globally categorize this fluent as either primary or secondary. ■

Our two switches which are connected by a spring (recall Fig. 2.5) show that influence information may induce cycles and even be symmetrical.

Example 2.5.2. Let $\mathcal{E} = \{\texttt{s}_1, \texttt{s}_2\}$ and $\mathcal{F} = \{\texttt{up}^1\}$. The two switches being mutually dependent, the correct influence information $\mathcal{I}$ is the following.

$$\{(\texttt{up}(\texttt{s}_1), \texttt{up}(\texttt{s}_2)), (\texttt{up}(\texttt{s}_2), \texttt{up}(\texttt{s}_1))\}$$ ■

Given correct information of potential influence, causal relationships are extractible from state constraints as follows. For the sake of clarity, let us first focus on quantifier-free, i.e., propositional fluent formulas. The general idea is investigating all potential ramifications and excluding those which do not respect the notion of influence. To this end, we consider all possible violations of a state constraint and formulate suitable causal relationships which help 'correct' this. Let us get more precise. Suppose C is a state constraint. First we construct the minimal conjunctive normal form (CNF, for short)[11] of this formula. Obviously, C is violated if and only if some conjunct

[11] A CNF of a (non-tautological, non-contradictory) formula C is some logical equivalent of the form $C_1 \wedge \ldots \wedge C_n$ $(n \geq 1)$ such that each C_i is of the form $\ell_{i1} \vee \ldots \vee \ell_{im_i}$ $(m_i \geq 1)$ with each ℓ_{ij} being a fluent literal. The *minimal* CNF is a CNF such that for each conjunct $C_i = \ell_{i1} \vee \ldots \vee \ell_{im_i}$ there is no strict subset $L \subsetneq \{\ell_{i1} \vee \ldots \vee \ell_{im_i}\}$ such that $C \models \bigvee_{\ell \in L} \ell$. The minimal CNF is unique modulo associativity, commutativity, and idempotency of $\wedge$ and $\vee$.

input C : state constraint;
$\mathcal{I}$: influence information;
output $\mathcal{R}$: set of causal relationships;
begin
  let $\mathcal{R} = \{\}$;
  let $C_1 \wedge \ldots \wedge C_n$ be the minimal CNF of C;
  for all $i = 1$ **to** n **do**
    let $\ell_1 \vee \ldots \vee \ell_m = C_i$;
    for all $j = 1$ **to** m **do**
      for all $k = 1$ **to** m, $k \neq j$ **do**
        if $(\|\ell_j\|, \|\ell_k\|) \in \mathcal{I}$ **then**
          let $\mathcal{R} = \mathcal{R} \cup \{\neg\ell_j \text{ \underline{causes} } \ell_k \text{ \underline{if} } \bigwedge_{\substack{l=1,\ldots,m \\ l\neq j, l\neq k}} \neg\ell_l\}$
  end-for
end

Figure 2.6. The automatic extraction of causal relationships from state constraints on the basis of influence information.

$\ell_1 \vee \ldots \vee \ell_m$ of its CNF is violated. This in turn means that $\neg\ell_1 \wedge \ldots \wedge \neg\ell_m$ holds. Since the initial state supposedly has satisfied constraint C, the reason for $\neg\ell_1 \wedge \ldots \wedge \neg\ell_m$ being true must be the occurrence of some (direct or indirect) effect $\neg\ell_j \in \{\neg\ell_1, \ldots, \neg\ell_m\}$. This violation of C can be 'corrected' by changing some other literal $\neg\ell_k$ of this collection to ℓ_k via a causal relationship—but only in case fluent $\|\ell_j\|$ potentially affects fluent $\|\ell_k\|$ according to $\mathcal{I}$. This way of generating causal relationships is formalized in the algorithm depicted in Fig. 2.6.

Example 2.5.3. Consider the entities $\mathcal{E} = \{\mathtt{s_1}, \mathtt{s_2}, \mathtt{s_3}\}$ along with the fluents $\mathcal{F} = \{\mathtt{up}^1, \mathtt{light}^0, \mathtt{relay}^0\}$ and influence information $\mathcal{I} = (2.9)$. Furthermore, let $\mathcal{C}$ be the familiar state constraints

$$\begin{aligned} \mathtt{light} &\equiv \mathtt{up(s_1)} \wedge \mathtt{up(s_2)} \\ \mathtt{relay} &\equiv \neg\mathtt{up(s_1)} \wedge \mathtt{up(s_3)} \\ \mathtt{relay} &\supset \neg\mathtt{up(s_2)} \end{aligned}$$

Our algorithm applied to the elements of $\mathcal{C}$ in conjunction with $\mathcal{I}$ traces as follows:

1. The minimal CNF of $\mathtt{light} \equiv \mathtt{up(s_1)} \wedge \mathtt{up(s_2)}$ is

   $$(\neg\mathtt{up(s_1)} \vee \neg\mathtt{up(s_2)} \vee \mathtt{light}) \wedge (\mathtt{up(s_1)} \vee \neg\mathtt{light}) \wedge (\mathtt{up(s_2)} \vee \neg\mathtt{light})$$

   Regarding conjunct $C_1 = \neg\mathtt{up(s_1)} \vee \neg\mathtt{up(s_2)} \vee \mathtt{light}$ we obtain the following:

– In case $j = 1, k = 2$ we have $(\texttt{up(s}_1\texttt{)}, \texttt{up(s}_2\texttt{)}) \notin \mathcal{I}$.
– In case $j = 1, k = 3$ we have $(\texttt{up(s}_1\texttt{)}, \texttt{light}) \in \mathcal{I}$, which yields

$$\texttt{up(s}_1\texttt{)} \ \underline{\text{causes}} \ \texttt{light} \ \underline{\text{if}} \ \texttt{up(s}_2\texttt{)}$$

– In case $j = 2, k = 1$ we have $(\texttt{up(s}_2\texttt{)}, \texttt{up(s}_1\texttt{)}) \notin \mathcal{I}$.
– In case $j = 2, k = 3$ we have $(\texttt{up(s}_2\texttt{)}, \texttt{light}) \in \mathcal{I}$, which yields

$$\texttt{up(s}_2\texttt{)} \ \underline{\text{causes}} \ \texttt{light} \ \underline{\text{if}} \ \texttt{up(s}_1\texttt{)}$$

– In case $j = 3, k = 1$ we have $(\texttt{light}, \texttt{up(s}_1\texttt{)}) \notin \mathcal{I}$.
– In case $j = 3, k = 2$ we have $(\texttt{light}, \texttt{up(s}_2\texttt{)}) \notin \mathcal{I}$.
Regarding conjunct $C_2 = \texttt{up(s}_1\texttt{)} \vee \neg\texttt{light}$ we obtain the following:
– In case $j = 1, k = 2$ we have $(\texttt{up(s}_1\texttt{)}, \texttt{light}) \in \mathcal{I}$, which yields

$$\neg\texttt{up(s}_1\texttt{)} \ \underline{\text{causes}} \ \neg\texttt{light} \ \underline{\text{if}} \ \top$$

– In case $j = 2, k = 1$ we have $(\texttt{light}, \texttt{up(s}_1\texttt{)}) \notin \mathcal{I}$.
Regarding conjunct $C_3 = \texttt{up(s}_2\texttt{)} \vee \neg\texttt{light}$ we obtain the following:
– In case $j = 1, k = 2$ we have $(\texttt{up(s}_2\texttt{)}, \texttt{light}) \in \mathcal{I}$, which yields

$$\neg\texttt{up(s}_2\texttt{)} \ \underline{\text{causes}} \ \neg\texttt{light} \ \underline{\text{if}} \ \top$$

– In case $j = 2, k = 1$ we have $(\texttt{light}, \texttt{up(s}_2\texttt{)}) \notin \mathcal{I}$.

2. The minimal CNF of $\texttt{relay} \equiv \neg\texttt{up(s}_1\texttt{)} \wedge \texttt{up(s}_3\texttt{)}$ is

$$(\texttt{up(s}_1\texttt{)} \vee \neg\texttt{up(s}_3\texttt{)} \vee \texttt{relay}) \wedge (\neg\texttt{up(s}_1\texttt{)} \vee \neg\texttt{relay}) \wedge (\texttt{up(s}_3\texttt{)} \vee \neg\texttt{relay})$$

Regarding conjunct $C_1 = \texttt{up(s}_1\texttt{)} \vee \neg\texttt{up(s}_3\texttt{)} \vee \texttt{relay}$ we obtain the following:
– In case $j = 1, k = 2$ we have $(\texttt{up(s}_1\texttt{)}, \texttt{up(s}_3\texttt{)}) \notin \mathcal{I}$.
– In case $j = 1, k = 3$ we have $(\texttt{up(s}_1\texttt{)}, \texttt{relay}) \in \mathcal{I}$, which yields

$$\neg\texttt{up(s}_1\texttt{)} \ \underline{\text{causes}} \ \texttt{relay} \ \underline{\text{if}} \ \texttt{up(s}_3\texttt{)}$$

– In case $j = 2, k = 1$ we have $(\texttt{up(s}_3\texttt{)}, \texttt{up(s}_1\texttt{)}) \notin \mathcal{I}$.
– In case $j = 2, k = 3$ we have $(\texttt{up(s}_3\texttt{)}, \texttt{relay}) \in \mathcal{I}$, which yields

$$\texttt{up(s}_3\texttt{)} \ \underline{\text{causes}} \ \texttt{relay} \ \underline{\text{if}} \ \neg\texttt{up(s}_1\texttt{)}$$

– In case $j = 3, k = 1$ we have $(\texttt{relay}, \texttt{up(s}_1\texttt{)}) \notin \mathcal{I}$.
– In case $j = 3, k = 2$ we have $(\texttt{relay}, \texttt{up(s}_3\texttt{)}) \notin \mathcal{I}$.
Regarding conjunct $C_2 = \neg\texttt{up(s}_1\texttt{)} \vee \neg\texttt{relay}$ we obtain the following:
– In case $j = 1, k = 2$ we have $(\texttt{up(s}_1\texttt{)}, \texttt{relay}) \in \mathcal{I}$, which yields

$$\texttt{up(s}_1\texttt{)} \ \underline{\text{causes}} \ \neg\texttt{relay} \ \underline{\text{if}} \ \top$$

– In case $j = 2, k = 1$ we have $(\texttt{relay}, \texttt{up(s}_1\texttt{)}) \notin \mathcal{I}$.
Regarding conjunct $C_3 = \texttt{up(s}_3\texttt{)} \vee \neg\texttt{relay}$ we obtain the following:

– In case $j = 1, k = 2$ we have $(\mathtt{up(s_3)}, \mathtt{relay}) \in \mathcal{I}$, which yields

$$\neg\mathtt{up(s_3)}\ \underline{\text{causes}}\ \neg\mathtt{relay}\ \underline{\text{if}}\ \top$$

– In case $j = 2, k = 1$ we have $(\mathtt{relay}, \mathtt{up(s_3)}) \notin \mathcal{I}$.

3. The minimal CNF of $\mathtt{relay} \supset \neg\mathtt{up(s_2)}$ is

$$\neg\mathtt{relay} \vee \neg\mathtt{up(s_2)}$$

Regarding the only conjunct $C_1 = \neg\mathtt{relay} \vee \neg\mathtt{up(s_2)}$ we obtain the following:

– In case $j = 1, k = 2$ we have $(\mathtt{relay}, \mathtt{up(s_2)}) \in \mathcal{I}$, which yields

$$\mathtt{relay}\ \underline{\text{causes}}\ \neg\mathtt{up(s_2)}\ \underline{\text{if}}\ \top$$

– In case $j = 2, k = 1$ we have $(\mathtt{up(s_2)}, \mathtt{relay}) \notin \mathcal{I}$.

Altogether, the output is exactly the nine causal relationships which we have obtained in the preceding section by intuitive analysis of causal dependencies. ■

As a small exercise, the reader may verify that the algorithm applied to the switches and spring-domain (Example 2.4.1 and Fig. 2.5) as well produces the correct outcome: The input consisting of the state constraint $\mathtt{up(s_1)} \equiv \mathtt{up(s_2)}$ and influence information $\mathcal{I} = \{(\mathtt{up(s_1)}, \mathtt{up(s_2)}), (\mathtt{up(s_2)}, \mathtt{up(s_1)})\}$ results in the output of all four expected causal relationships, (2.4).

The causal relationships ε causes ϱ if Φ generated by our procedure all have a restricted syntax. Namely, context Φ is a conjunction of fluent literals instead of an arbitrarily complex fluent formula. This does not imply, however, that some causal information otherwise being representable cannot be obtained through automatic generation. This is so, because any causal relationship can be transformed into an operationally equivalent set of causal relationships which obey the restriction. For suppose $\Phi_1 \vee \ldots \vee \Phi_n$ is a disjunctive normal form (DNF, for short)[12] of some formula Ψ, then a causal relationship ε causes ϱ if Ψ and the collection ε causes ϱ if Φ_1, ..., ε causes ϱ if Φ_n are interchangeable. On the other hand, in certain cases exploiting full expressiveness leads to considerably more compact representations. This can be accomplished by considering all automatically extracted causal relationships for a particular ε and ϱ, i.e., ε causes ϱ if Φ_1, ..., ε causes ϱ if Φ_n, and constructing formula Ψ as compact equivalent of $\Phi_1 \vee \ldots \vee \Phi_n$. Then ε causes ϱ if Ψ replaces the aforementioned n relationships. This first transforming state constraints into a normal form, then extracting causal relationships, and finally retransforming the result into non-normal form may of course require extensive computation. In certain cases it might be much faster to employ more sophisticated means to straightly extract causal relationships

[12] On the analogy of CNF, a DNF of a formula is some logical equivalent of the form $D_1 \vee \ldots \vee D_n$ $(n \geq 1)$ such that each D_i is of the form $\ell_{i1} \wedge \ldots \wedge \ell_{im_i}$ $(m_i \geq 1)$ with each ℓ_{ij} being a fluent literal.

from state constraints, i.e., by avoiding the roundabout way of constructing normal forms. On the other hand, the generation of causal relationships is a pre-processing step which is carried out only once for a fixed set of state constraints. This computational effort is therefore of minor importance.

Talking about complexity, a related crucial issue concerns the number of causal relationships generated from a set of state constraints. This number is, in the worst case, exponential as regards the size of the input. The reason is that formulas exist which admit only CNF's of exponentially increased length. Moreover, up to quadratic many relationships exist for a single conjunct, namely, if all involved fluents have the potential to affect each other. Despite this negative result, fortunately there is a decisive characteristic due to which especially in large domains the number of relationships is small compared to the worst case: The state constraints do not interfere when determining causal relationships. In general, large domains tend to be locally structured in that each single state constraint relates only a small fraction of the entire set of fluents. More precisely, let k be the maximum size of a state constraint[13] and n their overall number. Then the number of required causal relationships is of magnitude $O(2^k \cdot n)$. Granted that k remains constant with increasing domain size, the number of causal relationships thus is linear in the number of state constraints. E.g., in our example involving n sub-circuits (recall Fig. 2.2), each of the n state constraints relates only three fluents. Consequently, as we have seen (c.f. (2.8)) they give rise to a linear number of causal relationships ($4 \cdot n$, to be precise). Incidentally, the fact that state constraints do not interfere in determining causal relationships solves the second problem mentioned in the introduction to the Ramification Problem. No existing causal relationship requires modification or needs to be removed if state constraints are added. Introducing a new switch-bulb pair $\mathtt{up}(\mathtt{s}_{n+1}), \mathtt{light}_{n+1}$ in the example just mentioned, for instance, amounts to adding four new causal relationships but requires no further changes.

Thus far we have confined ourselves to the extraction of causal relationships from quantifier-free state constraints. As for the general case, notice first that since we assume finiteness of sets of entities $\mathcal{E}$, it is always possible to rewrite any state constraint so as to become quantifier-free. Namely, each sub-formula $\forall x. F$ is replaced by $\bigwedge_{e \in \mathcal{E}} F\{x \mapsto e\}$, and each sub-formula $\exists x. F$ is replaced by $\bigvee_{e \in \mathcal{E}} F\{x \mapsto e\}$. Awkward as it is, this is unavoidable when dealing with unrestricted state constraints and influence information. To see why, consider some arbitrary formula like $\forall x \exists y \forall z. f(x, y, z)$ as state constraint. Which causal relationships are determined by this constraints depends on how the various possible instances $f(e_1, e_2, e_3)$ interact. Each of the instances may behave differently with this respect. In general, therefore, there is no better way than grounding a state constraint and proceeding as above. On the other hand, it is possible to exploit the expressiveness of causal relationships having variables in case a state constraint and influence information obey well-defined restrictions. In the following, we discuss two such

[13] As the size of a formula we take the number of literal occurrences.

classes which we consider of general importance so that a lot is gained by their special treatment.

First, consider state constraints of the form $\forall x_1 \ldots \forall x_n. F$ (or $\forall \overline{x}. F$, for short) with sub-formula F being quantifier-free. Suppose further that for any atomic fluent expression $f[\overline{x}]$ occurring in F, the underlying influence information does not differ regarding different instances $f[\overline{e}]$. Then the causal relationships determined by this constraint and influence information can be obtained by applying our procedure to sub-formula F with all occurrences of variables taken as (distinct) entities.[14] As a simple example, consider state constraint $\forall x\, [\, \mathtt{down}(x) \equiv \neg \mathtt{up}(x)\,]$ and influence information $\mathcal{I} = \{(\mathtt{up}(x), \mathtt{down}(x)), (\mathtt{down}(x), \mathtt{up}(x))\}$. Four causal relationships are extractible from the corresponding quantifier-free constraint, $\mathtt{down}(x) \equiv \neg \mathtt{up}(x)$, viz.

$$
\begin{array}{rcrcl}
\mathtt{down}(x) & \underline{\text{causes}} & \neg\mathtt{up}(x) & \underline{\text{if}} & \top \\
\neg\mathtt{down}(x) & \underline{\text{causes}} & \mathtt{up}(x) & \underline{\text{if}} & \top \\
\mathtt{up}(x) & \underline{\text{causes}} & \neg\mathtt{down}(x) & \underline{\text{if}} & \top \\
\neg\mathtt{up}(x) & \underline{\text{causes}} & \mathtt{down}(x) & \underline{\text{if}} & \top
\end{array}
$$

stating that any entity becoming up (or not up) also becomes not down (or down, respectively) and vice versa.

Second, and more sophisticated, we analyze state constraints in which each occurrence of a quantifier is of the form $\forall \overline{x}. \ell[\overline{x}]$ or of the form $\exists \overline{x}. \ell[\overline{x}]$ where $\ell[\overline{x}]$ is a (possibly negated) fluent expression with free variables $\overline{x}$. Furthermore, it is assumed that, for any such $\ell[\overline{x}]$, the underlying influence information does not differ regarding different instances $\|\ell[\overline{e}]\|$. When computing the minimal CNF of a state constraint which is restricted in this way, sub-formulas $\forall \overline{x}. \ell[\overline{x}]$ and $\exists \overline{x}. \ell[\overline{x}]$ are treated as if they were ordinary literals. If in addition any $\neg \forall \overline{x}. \ell[\overline{x}]$ is replaced by $\exists \overline{x}. \neg \ell[\overline{x}]$ and any $\neg \exists \overline{x}. \ell[\overline{x}]$ by $\forall \overline{x}. \neg \ell[\overline{x}]$, then each conjunct in the resulting CNF is a disjunction consisting of ground literals and expressions of the form $\forall \overline{x}. \ell[\overline{x}]$ and $\exists \overline{x}. \ell[\overline{x}]$. When generating causal relationships, the sub-formulas with quantifier are treated as follows: The reason for a formula $\forall \overline{x}. \ell[\overline{x}]$ becoming false must be the occurrence of any effect $\neg \ell[\overline{e}]$. Thus $\neg \ell[\overline{x}]$ itself may be used, if influence permits, as triggering effect in a causal relationship. The reason for a formula $\exists \overline{x}. \ell[\overline{x}]$ becoming false must be the occurrence of any effect $\neg \ell[\overline{e}]$ so that (now) $\forall \overline{x}. \neg \ell[\overline{x}]$ holds. Thus $\neg \ell[\overline{x}]$ may be used as triggering effect in a causal relationship whose context includes $\forall \overline{x}. \neg \ell[\overline{x}]$. Correcting the violation of a state constraint by satisfying a formula $\forall \overline{x}. \ell[\overline{x}]$ is achieved by satisfying all instances of $\ell[\overline{x}]$. Thus $\ell[\overline{x}]$ may be used as indirect effect in a causal relationship. Correcting the violation of a state constraint by satisfying a formula $\exists \overline{x}. \ell[\overline{x}]$ is achieved by satisfying just one instance of $\ell[\overline{x}]$. Thus $\ell[\overline{x}]$

[14] During this procedure, the underlying influence information should of course be assumed to treat these entities similar to all others as regards the aforementioned atomic fluent expressions $f[\overline{x}]$ contained in F.

input C : state constraint;
$\mathcal{I}$: influence information;
output $\mathcal{R}$: set of causal relationships;
begin
  let $\mathcal{R} = \{\}$;
  let $C_1 \wedge \ldots \wedge C_n$ be the minimal CNF of C ;
  for all $i = 1$ **to** n **do**
    let $F_1 \vee \ldots \vee F_m = C_i$;
    for all $j = 1$ **to** m **do**
      for all $k = 1$ **to** $m,\ k \neq j$ **do**
        let $\Phi = \bigwedge_{l = 1, \ldots, m;\ l \neq j, l \neq k} \neg F_l$;
        case F_j **is** ℓ_j**:**
          case F_k **is** ℓ_k**: let** $r = \neg\ell_j$ <u>causes</u> ℓ_k <u>if</u> Φ;
          case F_k **is** $\forall\overline{x}_k.\,\ell_k[\overline{x}_k]$**: let** $r = \neg\ell_j$ <u>causes</u> $\ell_k[\overline{x}_k]$ <u>if</u> Φ;
          case F_k **is** $\exists\overline{x}_k.\,\ell_k[\overline{x}_k]$**:**
            let $r = \neg\ell_j$ <u>causes</u> $\ell_k[\overline{x}_k]$ <u>if</u> $\forall\overline{x}_k.\,\neg\ell_k[\overline{x}_k] \wedge \Phi$
        case F_j **is** $\forall\overline{x}_j.\,\ell_j[\overline{x}_j]$**:**
          case F_k **is** ℓ_k**: let** $r = \neg\ell_j[\overline{x}_j]$ <u>causes</u> ℓ_k <u>if</u> Φ;
          case F_k **is** $\forall\overline{x}_k.\,\ell_k[\overline{x}_k]$**: let** $r = \neg\ell_j[\overline{x}_j]$ <u>causes</u> $\ell_k[\overline{x}_k]$ <u>if</u> Φ;
          case F_k **is** $\exists\overline{x}_k.\,\ell_k[\overline{x}_k]$**:**
            let $r = \neg\ell_j[\overline{x}_j]$ <u>causes</u> $\ell_k[\overline{x}_k]$ <u>if</u> $\forall\overline{x}_k.\,\neg\ell_k[\overline{x}_k] \wedge \Phi$
        case F_j **is** $\exists\overline{x}_j.\,\ell_j[\overline{x}_j]$**:**
          case F_k **is** ℓ_k**: let** $r = \neg\ell_j[\overline{x}_j]$ <u>causes</u> ℓ_k <u>if</u> $\forall\overline{x}_j.\,\neg\ell_j[\overline{x}_j] \wedge \Phi$;
          case F_k **is** $\forall\overline{x}_k.\,\ell_k[\overline{x}_k]$**:**
            let $r = \neg\ell_j[\overline{x}_j]$ <u>causes</u> $\ell_k[\overline{x}_k]$ <u>if</u> $\forall\overline{x}_j.\,\neg\ell_j[\overline{x}_j] \wedge \Phi$;
          case F_k **is** $\exists\overline{x}_k.\,\ell_k[\overline{x}_k]$**:**
            let $r = \neg\ell_j[\overline{x}_j]$ <u>causes</u> $\ell_k[\overline{x}_k]$ <u>if</u> $\forall\overline{x}_j.\,\neg\ell_j[\overline{x}_j] \wedge \forall\overline{x}_k.\,\neg\ell_k[\overline{x}_k] \wedge \Phi$
        if $(\|\ell_j[\overline{x}_j]\|, \|\ell_k[\overline{x}_k]\|) \in \mathcal{I}$ **then**
          let $\mathcal{R} = \mathcal{R} \cup \{r\}$
      end-for
    end-for
end

Figure 2.7. An extended procedure for the automatic extraction of causal relationships.

may be used as indirect effect in a causal relationship whose context includes $\forall\overline{x}.\,\neg\ell[\overline{x}]$. On this basis, an extended algorithm for automatic generation of causal relationships is depicted in Fig. 2.7.

Example 2.5.4. Consider state constraint $C = \mathtt{light} \equiv \forall x.\,\mathtt{up}(x)$, expressing the serial connection of all involved switches and a light bulb. Let $\mathcal{I} = \{(\mathtt{up}(x), \mathtt{light})\}$, then our extended algorithm traces as follows:

The minimal CNF of C is

$$(\exists x. \neg \texttt{up}(x) \vee \texttt{light}) \wedge (\forall y. \texttt{up}(y) \vee \neg \texttt{light})$$

1. Regarding conjunct $C_1 = \exists x. \neg \texttt{up}(x) \vee \texttt{light}$ we obtain the following:
   - In case $j = 1, k = 2$ we have $(\texttt{up}(x), \texttt{light}) \in \mathcal{I}$, which yields

$$\texttt{up}(x) \ \underline{\text{causes}} \ \texttt{light} \ \underline{\text{if}} \ \forall x. \texttt{up}(x)$$

   - In case $j = 2, k = 1$ we have $(\texttt{light}, \texttt{up}(x)) \notin \mathcal{I}$.
2. Regarding conjunct $C_2 = \forall y. \texttt{up}(y) \vee \neg \texttt{light}$ we obtain the following:
   - In case $j = 1, k = 2$ we have $(\texttt{up}(y), \texttt{light}) \in \mathcal{I}$, which yields

$$\neg \texttt{up}(y) \ \underline{\text{causes}} \ \neg \texttt{light} \ \underline{\text{if}} \ \top$$

   - In case $j = 2, k = 1$ we have $(\texttt{light}, \texttt{up}(y)) \notin \mathcal{I}$.

Thus closing either switch causes light provided all other switches are closed; and conversely, opening either switch causes the light be off afterwards regardless of other switches—as one would expect. ■

With this extension of our basic algorithm we conclude the discussion on how causal relationships may be automatically generated whenever suitable influence information can be provided. It is, however, not claimed that the latter is always possible with our basic notion of influence information. More sophisticated means to specify potential influence may be required in certain domains. E.g., a fluent having the potential to affect another fluent may depend on whether the former occurs affirmatively or negated. To reflect this, the concept of influence information needs to be extended so as to relate fluent literals and not just fluents. Further generalization may allow for restricting potential influence to circumstances expressed by arbitrary fluent formulas.[15] Whichever sophistication is required depends on the domain at hand. The simpler the notion of potential influence—provided it suffices—, the more is gained by the automatic extraction of causal relationships compared to these drawing up all by hand.

2.6 Non-minimal Successor States

The attempts to solve the Ramification Problem which we have discussed prior to causal relationships were based on the idea of minimizing change to the largest reasonable extent. The employment of causal relationships does

[15] Notice that this is unnecessary in case the circumstances follow from the state constraints. Of course, fluent `relay` can affect $\texttt{up}(\texttt{s}_2)$ only in case `relay` is true (in our circuit depicted in Fig. 2.4). However, the very state constraint $\texttt{relay} \supset \neg \texttt{up}(\texttt{s}_2)$ encodes this knowledge, which is why it needs not be integrated into the information of potential influence.

not, at least not *a priori*, respect a similar notion. Since the principle of minimizing change is widely considered essential for the Ramification Problem in literature, the present section is devoted to an analysis of whether and how this principle and the successive application of causal relationships relate. It will turn out that causal successor states cover all states with minimal distance to some preliminary successors in satisfying the state constraints while respecting causality. Yet we will also show that the converse does not hold. That is, requiring minimality is proved failing to account for all possible causal successor states in some domains. This result challenges the common belief in the necessity of minimizing change when addressing the Ramification Problem.

For a general solution to the Ramification Problem, as we have seen all along, mere state constraints contain insufficient information. A suitable minimization strategy therefore requires additional domain knowledge. The purpose of this knowledge is to help telling apart derived implications which correspond to actual indirect effects from those which are just logical consequences. For example, $\mathtt{up(s_1)} \land \mathtt{up(s_2)} \supset \mathtt{light}$, derived from state constraint $\mathtt{light} \equiv \mathtt{up(s_1)} \land \mathtt{up(s_2)}$, indicates a causally correct implication, whereas $\mathtt{up(s_1)} \land \neg\mathtt{light} \supset \neg\mathtt{up(s_2)}$, derived from the same constraint, does not. Though being logically equivalent, the two implications need to be distinguished when minimizing change while accounting for indirect effects. Causal relationships serve this purpose if taken as directed implications. Unlike classical material implications, these directed implications must not be applied in reverse direction. That is, while the two fluent formulas $F \land \ell_1 \supset \ell_2$ and $F \land \neg\ell_2 \supset \neg\ell_1$ are interchangeable, the semantics of the corresponding relationships ℓ_1 <u>causes</u> ℓ_2 <u>if</u> F and $\neg\ell_2$ <u>causes</u> $\neg\ell_1$ <u>if</u> F shall differ also in their logical reading. This reading of causal relationships as directed implications shall be the basis for a formal definition of minimizing change with regard to causal knowledge. This first of all requires a notion of how to employ causal relationships as deduction rules. Similar to the operational reading of causal relationships, this requires to distinguish between a context and a set of actually occurred effects. For the purpose of deduction, the former is given as a set of arbitrary fluent formulas Ψ while the latter is, as before, a set of fluent literals, Θ. In what follows, if F is a fluent formula, then we say that a set of fluent formulas Ψ *entails* F if F is true in all states satisfying Ψ. By $Th(\Psi)$ we denote the set of all formulas thus entailed by Ψ (that is, the *theory* of Ψ). The derivable consequences of a context-effect pair given underlying causal relationships is then defined as follows.

Definition 2.6.1. *Let $\mathcal{E}$ and $\mathcal{F}$ be sets of entities and fluent names, respectively, and $\mathcal{R}$ a set of causal relationships. If Ψ is a set of fluent formulas and Θ a set of fluent literals, then the* theory *induced by (Ψ, Θ) relative to $\mathcal{R}$, written $Th_{\mathcal{R}}(\Psi, \Theta)$, is the smallest (wrt. set inclusion) pair $(\widehat{\Psi}, \hat{\Theta})$ such that*

1. $\Psi \subseteq \widehat{\Psi}$ *and* $\Theta \subseteq \hat{\Theta}$;
2. $\widehat{\Psi} = Th(\widehat{\Psi})$; *and*
3. *for any instance* ε <u>causes</u> ϱ <u>if</u> $\Phi \in \mathcal{R}$ *such that* $\Phi \in \widehat{\Psi}$ *and* $\varepsilon \in \hat{\Theta}$, *we have* $\varrho \in \widehat{\Psi}$ *and* $\varrho \in \hat{\Theta}$.

If $Th_{\mathcal{R}}(\Psi, \Theta) = (\widehat{\Psi}, \hat{\Theta})$ *and* $F \in \widehat{\Psi}$, *then this is denoted by* $(\Psi, \Theta) \Vdash_{\mathcal{R}} F$. ∎

It is easy to see that $\Vdash_{\mathcal{R}}$ is monotonic, that is, $\Psi_1 \subseteq \Psi_2$ and $\Theta_1 \subseteq \Theta_2$ implies $\{F : (\Psi_1, \Theta_1) \Vdash_{\mathcal{R}} F\} \subseteq \{F : (\Psi_2, \Theta_2) \Vdash_{\mathcal{R}} F\}$.

Example 2.6.1. Let $\mathcal{E} = \{\mathtt{s_1}, \mathtt{s_2}\}$ and $\mathcal{F} = \{\mathtt{up}^1, \mathtt{light}^0\}$, and let $\mathcal{R}$ consist of the causal relationships reflecting the causal dependencies between the two switches and the light bulb in Fig. 2.3, i.e.,

$$\begin{array}{ll} \mathtt{up(s_1)}\ \underline{\text{causes}}\ \mathtt{light}\ \underline{\text{if}}\ \mathtt{up(s_2)} & \neg\mathtt{up(s_1)}\ \underline{\text{causes}}\ \neg\mathtt{light}\ \underline{\text{if}}\ \top \\ \mathtt{up(s_2)}\ \underline{\text{causes}}\ \mathtt{light}\ \underline{\text{if}}\ \mathtt{up(s_1)} & \neg\mathtt{up(s_2)}\ \underline{\text{causes}}\ \neg\mathtt{light}\ \underline{\text{if}}\ \top \end{array} \tag{2.10}$$

Now, consider the set $\Psi_1 = \{\mathtt{up(s_1)} \wedge \mathtt{up(s_2)}\}$ along with $\Theta = \{\mathtt{up(s_1)}\}$. Then $Th_{\mathcal{R}}(\Psi_1, \Theta) = (Th(\{\mathtt{up(s_1)} \wedge \mathtt{up(s_2)}, \mathtt{light}\}), \{\mathtt{up(s_1)}, \mathtt{light}\})$; hence, $(\Psi_1, \Theta) \Vdash_{\mathcal{R}} \mathtt{light}$. In contrast, suppose $\Psi_2 = \{\mathtt{up(s_1)} \wedge \neg\mathtt{light}\}$ and, as before, $\Theta = \{\mathtt{up(s_1)}\}$, then $Th_{\mathcal{R}}(\Psi_2, \Theta) = (Th(\{\mathtt{up(s_1)} \wedge \neg\mathtt{light}\}), \{\mathtt{up(s_1)}\})$ since no causal relationship is applicable. Hence, $(\Psi_2, \Theta) \nVdash_{\mathcal{R}} \neg\mathtt{up(s_2)}$. ∎

The notion of performing deduction on the basis of causal relationships shall be exploited for the construction of a fixpoint characterization of successor states which accounts for indirect effects while respecting causal information. Informally speaking, suppose an action with direct effect E is performed in state S. Then a state T is considered successor iff the following holds. Set T includes E, T along with E and the underlying causal relationships do not entail an inconsistency, and each fluent change from S to T is grounded on some causal relationship. This last condition reflects the idea of minimizing change.

Definition 2.6.2. *Let* $(\mathcal{E}, \mathcal{F}, \mathcal{A}, \mathcal{L})$ *be a basic action domain and* $\mathcal{R}$ *a set of causal relationships. If* S *is a state and* a *an action, then a state* T *is* causal minimizing-change successor *of* S *and* a *iff the following holds: Set* $\mathcal{L}$ *contains an applicable action law instance* a <u>transforms</u> C <u>into</u> E *such that*

$$T = \{\ell : ((S \cap T) \cup E, E) \Vdash_{\mathcal{R}} \ell\} \tag{2.11}$$

that is, T *is fixpoint of the function* $\lambda T. \{\ell : ((S \cap T) \cup E, E) \Vdash_{\mathcal{R}} \ell\}$ *given* S *and* E. ∎

Example 2.6.2. Let $\mathcal{D}$ be the basic action domain which models our circuit composed of two switches and a light bulb of Fig. 2.3 as in Example 2.2.2. Furthermore, let $\mathcal{R}$ consist in the four causal relationships (2.10) above. Suppose that the current state be $S = \{\neg\mathtt{up(s_1)}, \mathtt{up(s_2)}, \neg\mathtt{light}\}$ as depicted in Fig. 2.3. The only applicable action law instance for action $a = \mathtt{toggle(s_1)}$ is $\mathtt{toggle(s_1)}$ <u>transforms</u> $\{\neg\mathtt{up(s_1)}\}$ <u>into</u> $\{\mathtt{up(s_1)}\}$

with effect $E = \{\mathtt{up(s_1)}\}$. Then the only causal minimizing-change successor of S and a is $T = \{\mathtt{up(s_1)}, \mathtt{up(s_2)}, \mathtt{light}\}$. For $((S \cap T) \cup E, E) = (\{\mathtt{up(s_2)}, \mathtt{up(s_1)}\}, \{\mathtt{up(s_1)}\})$ allows to derive $\mathtt{light}$. In contrast, the unintended $T' = \{\mathtt{up(s_1)}, \neg\mathtt{up(s_2)}, \neg\mathtt{light}\}$ does not satisfy Equation (2.11). For $((S \cap T') \cup E, E) = (\{\neg\mathtt{light}, \mathtt{up(s_1)}\}, \{\mathtt{up(s_1)}\})$, which does not allow for deriving $\neg\mathtt{up(s_2)}$, i.e., the literal which is missing to regain T'.[16] ■

The following observation justifies formally our name "minimizing-change successor." Namely, any state T satisfying Equation (2.11) has minimal distance from S in that there is no state T' with less (wrt. set inclusion) changes while also satisfying this equation.

Observation 2.6.1. *Let $\mathcal{E}$ and $\mathcal{F}$ be sets of entities and fluent names, $\mathcal{R}$ a set of causal relationships, S a state, and E a consistent set of fluent literals. For any two fixpoints T, T' of $\lambda T. \{\ell : ((S \cap T) \cup E, E) \Vdash_{\mathcal{R}} \ell\}$, if $T \setminus S \supseteq T' \setminus S$, then $T = T'$.*

Proof. Since each of S, T, T' is a state, $T \setminus S = T' \setminus S$ implies $T = T'$. The assumption $T \setminus S \supsetneq T' \setminus S$ leads to the following contradiction. Let $\ell \in T \setminus S$ such that $\ell \notin T' \setminus S$, then $\ell \in T$ and $\ell \notin S$, hence $\neg\ell \in S$ and $\neg\ell \in T'$ as both S and T' are states. Since T' is consistent and fixpoint, we have $((S \cap T') \cup E, E) \not\Vdash_{\mathcal{R}} \ell$ due to $\neg\ell \in T'$. On the other hand, we know that $((S \cap T) \cup E, E) \Vdash_{\mathcal{R}} \ell$ due to $\ell \in T$. Operator $\Vdash_{\mathcal{R}}$ being monotonic, $((S \cap T') \cup E, E) \not\Vdash_{\mathcal{R}} \ell$ and $((S \cap T) \cup E, E) \Vdash_{\mathcal{R}} \ell$ together imply $S \cap T \not\subseteq S \cap T'$. Thus we can find some $\ell' \in S, T$ such that $\ell' \notin T'$. Then $\neg\ell' \notin S, T$ and $\neg\ell' \in T'$, which contradicts premise $T \setminus S \supseteq T' \setminus S$. *Qed.*

Before we enter the analysis of the relation between minimization based on causal relationships on the one hand, and their operating on preliminary successors on the other hand, we present an iterative characterization of theories induced by causal relationships. This alternative view will be helpful for later purpose.

Proposition 2.6.1. *Let $\mathcal{E}$ and $\mathcal{F}$ be sets of entities and fluent names, respectively, and $\mathcal{R}$ a set of causal relationships. For each pair (Ψ, Θ) consisting of a set of fluent formulas and a set of fluent literals, we define*

1. $\Psi_0 = \Psi$ *and* $\Theta_0 = \Theta$.
2. *For* $i > 0$,
   a) $\Psi_i = Th(\Psi_{i-1}) \cup \{\varrho : \varepsilon \ \underline{\text{causes}}\ \varrho \ \underline{\text{if}}\ \Phi \in \mathcal{R}, \Phi \in \Psi_{i-1}, \varrho \notin \Psi_{i-1}, \varepsilon \in \Theta_{i-1}\}$
   b) $\Theta_i = \Theta_{i-1} \cup \{\varrho : \varepsilon \ \underline{\text{causes}}\ \varrho \ \underline{\text{if}}\ \Phi \in \mathcal{R}, \Phi \in \Psi_{i-1}, \varrho \notin \Psi_{i-1}, \varepsilon \in \Theta_{i-1}\}$

Then $\widehat{\Psi} = \bigcup_{i=0}^{\infty} \Psi_i$ and $\hat{\Theta} = \bigcup_{i=0}^{\infty} \Theta_i$, where $(\widehat{\Psi}, \hat{\Theta}) = Th_{\mathcal{R}}(\Psi, \Theta)$.

[16] It is also interesting to see why $T'' = \{\mathtt{up(s_1)}, \mathtt{up(s_2)}, \neg\mathtt{light}\}$, where only the direct effect is computed, does not satisfy (2.11): Pair $((S \cap T'') \cup E, E) = (\{\mathtt{up(s_2)}, \neg\mathtt{light}, \mathtt{up(s_1)}\}, \{\mathtt{up(s_1)}\})$ additionally entails $\mathtt{light}$ with the help of $\mathcal{R}$. Thus, T'' is not a fixpoint. This illustrates that all material implications $\Phi \wedge \varepsilon \supset \varrho$ induced by some causal relationship ε causes ϱ if Φ hold in minimizing-change successors.

Proof.

1. We first prove $\widehat{\Psi} \subseteq \bigcup_{i=0}^{\infty} \Psi_i$ and $\hat{\Theta} \subseteq \bigcup_{i=0}^{\infty} \Theta_i$.
   a) Let $F \in \widehat{\Psi}$. Then minimality of $(\widehat{\Psi}, \hat{\Theta})$ implies the existence of a finite sequence ε_1 causes ϱ_1 if $\Phi_1, \ldots, \varepsilon_n$ causes ϱ_n if Φ_n of causal relationships in $\mathcal{R}$ such that $F \in Th(\Psi \cup \{\varrho_1, \ldots, \varrho_n\})$ and, for each $1 \leq i \leq n$, we have $\Phi_i \in Th(\Phi \cup \{\varrho_1, \ldots, \varrho_{i-1}\})$, $\varrho_i \notin Th(\Phi \cup \{\varrho_1, \ldots, \varrho_{i-1}\})$ (because of minimality of $\hat{\Theta}$) and $\varepsilon_i \in \Theta \cup \{\varrho_1, \ldots, \varrho_{i-1}\}$. Consequently, $F \in Th(\Psi_{n+1}) \subseteq \bigcup_{i=0}^{\infty} \Psi_i$.
   b) Let $\ell \in \hat{\Theta}$. An argument similar to the one in clause (a) ensures that $\ell \in \Theta_n \subseteq \bigcup_{i=0}^{\infty} \Theta_i$ for some $n \in \mathbb{N}_0$.
2. To show $\widehat{\Psi} \supseteq \bigcup_{i=0}^{\infty} \Psi_i$ and $\hat{\Theta} \supseteq \bigcup_{i=0}^{\infty} \Theta_i$, we prove by induction that $\Psi_n \subseteq \widehat{\Psi}$ and $\Theta_n \subseteq \hat{\Theta}$ for all $n \in \mathbb{N}_0$. The base case, $n = 0$, holds by definition since $\Psi_0 = \Psi \subseteq \widehat{\Psi}$ and $\Theta_0 = \Theta \subseteq \hat{\Theta}$. Suppose $n > 0$ and assume that the claim holds for $n - 1$.
   a) If $F \in Th(\Psi_{n-1})$, then the induction hypothesis $\Psi_{n-1} \subseteq \widehat{\Psi}$ implies $F \in Th(\widehat{\Psi}) = \widehat{\Psi}$.
   b) If $F = \varrho$ for some ε causes ϱ if $\Phi \in \mathcal{R}$ such that $\Phi \in \Psi_{n-1}$ and $\varepsilon \in \Theta_{n-1}$, then the induction hypothesis $\Psi_{n-1} \subseteq \widehat{\Psi}$ and $\Theta_{n-1} \subseteq \hat{\Theta}$, respectively, implies $\Phi \in \widehat{\Psi}$ and $\varepsilon \in \hat{\Theta}$. Consequently, $\varrho \in \widehat{\Psi}$ and $\varrho \in \hat{\Theta}$.

Qed.

We are now prepared for a formal proof of the announced subsumption result. That is, we will show that the successive application of causal relationships to preliminary successors enables us to encounter all causal minimizing-change successors.

Theorem 2.6.1. *Let $(\mathcal{E}, \mathcal{F}, \mathcal{A}, \mathcal{L})$ be a basic action domain and $\mathcal{C}$ and $\mathcal{R}$ be sets of state constraints and causal relationships, respectively. Furthermore, let S be an acceptable state and a an action. Then each causal minimizing-change successor of S and a is also causal successor state.*

Proof. Let T be a causal minimizing-change successor obtained through action law a transforms C into $E \in \mathcal{L}$. Let $\Psi = (S \cap T) \cup E$ and $\Theta = E$, then $T = \{\ell : ((S \cap T) \cup E, E) \Vdash_{\mathcal{R}} \ell\}$ implies $Th(T) = \widehat{\Psi} = \bigcup_{i=0}^{\infty} \Psi_i$ where $(\widehat{\Psi}, \hat{\Theta}) = Th_{\mathcal{R}}(\Psi, \Theta)$.

We prove by induction that, for each $n \in \mathbb{N}_0$, $((S \setminus C) \cup E, E) \overset{*}{\leadsto}_{\mathcal{R}} (S_n, E_n)$ for some (S_n, E_n) such that $\Psi_n \subseteq Th(S_n)$ and $\Theta_n = E_n$. Notice that each Ψ_n $(n \geq 0)$ is consistent as $\Psi_n \subseteq Th(T)$ and T is a state.

In case $n = 0$, let $S_0 = (S \setminus C) \cup E$ and $E_0 = E$, which together satisfy the claim: Consistency of $\Psi_0 = (S \cap T) \cup E$ and $(S \setminus C) \cup E$ being a state implies $(S \cap T) \cup E \subseteq (S \setminus C) \cup E$, hence $\Psi_0 \subseteq S_0$. Set E_0 equals Θ_0 by definition.

For the induction step, let $n > 0$ and suppose that some (S_{n-1}, E_{n-1}) satisfies the claim. Let

$$\{\varrho_1, \ldots, \varrho_m\} = \Psi_n \setminus \Psi_{n-1} \tag{2.12}$$

be the set of all fluent literals that are added to Ψ_{n-1} to obtain Ψ_n. Then there exist m causal relationships ε_1 <u>causes</u> ϱ_1 <u>if</u> $\Phi_1, \ldots, \varepsilon_m$ <u>causes</u> ϱ_m <u>if</u> Φ_m such that $\Phi_j \in \Psi_{n-1}$, $\varrho_j \notin \Psi_{n-1}$, and $\varepsilon_j \in \Theta_{n-1}$ for each $1 \leq j \leq m$. Consider the first relationship, i.e., $r_1 = \varepsilon_1$ <u>causes</u> ϱ_1 <u>if</u> Φ_1. According to the induction hypothesis, we know $\Psi_{n-1} \subseteq Th(S_{n-1})$ and $\Theta_{n-1} = E_{n-1}$. Consequently, Φ_1 is true in S_{n-1} and $\varepsilon_1 \in E_{n-1}$. Moreover, we have that $\neg\varrho_1 \in S_{n-1}$, for assuming the contrary leads to the following contradiction. Given that $\varrho_1 \in S_{n-1}$, in case $\varrho_1 \in E_{n-1}$ we find that $\varrho_1 \in \Psi_{n-1}$ due to $E_{n-1} = \Theta_{n-1}$ and $\Theta_{n-1} \subseteq \Psi_{n-1}$—which contradicts $\varrho_1 \notin \Psi_{n-1}$ according to Equation (2.12). In case $\varrho_1 \notin E_{n-1}$, we have $\varrho_1 \in S_{n-1}$ implies $\varrho_1 \in S$. From $\varrho_1 \in \Psi_n \subseteq \bigcup_{i=0}^{\infty} \Psi_i = T$ it follows that $\varrho_1 \in (S \cap T) \subseteq \Psi_0 \subseteq \Psi_{n-1}$—which again contradicts $\varrho_1 \notin \Psi_{n-1}$ according to Equation (2.12).

Thus we can apply relationship r_1 to the state-effect pair (S_{n-1}, E_{n-1}). Consider, now, the second causal relationship $r_2 = \varepsilon_2$ <u>causes</u> ϱ_2 <u>if</u> Φ_2. Just like r_1, relationship r_2 is applicable to (S_{n-1}, E_{n-1}). We have to prove that it is still applicable after having applied r_1. Condition $\varepsilon_2 \in E_{n-1}$ does not become violated, for otherwise we had $\varepsilon_2 = \neg\varrho_1$, which would imply $\neg\varrho_1 \in E_{n-1} = \Theta_{n-1} \subseteq \Psi_{n-1} \subseteq \Psi_n$, which contradicts $\varrho_1 \in \Psi_n$ according to Equation (2.12). Condition $\neg\varrho_2 \in S_{n-1}$ as well does not become violated, for otherwise we had $\varrho_1 = \varrho_2$, which contradicts the left hand side of Equation (2.12) being a set. Finally, condition $\Phi_2 \in Th(S_{n-1})$ does not become violated because $\Phi_2 \in \Psi_{n-1}$ and $\Psi_{n-1} \subseteq Th(S_{n-1})$ imply $\Phi_2 \in Th((S_{n-1} \setminus \{\neg\varrho_1\}) \cup \{\varrho_1\})$ since Ψ_{n-1} does not contain ϱ_1 nor $\neg\varrho_1$. Thus we can successively apply all m causal relationships to (S_{n-1}, E_{n-1}). The overall resulting state-effect pair (S_n, E_n), where

$$\begin{aligned} S_n &= (S_{n-1} \setminus \{\neg\varrho_1, \ldots, \neg\varrho_m\}) \cup \{\varrho_1, \ldots, \varrho_m\} \\ E_n &= E_{n-1} \cup \{\varrho_1, \ldots, \varrho_m\} \end{aligned}$$

satisfies the claim because $\Psi_n \subseteq Th(S_n)$ and $\Theta_n = E_n$.

Now, since there exists only a finite number of changes from S to T, we have $T = \{\ell : \ell \in \bigcup_{i=0}^{\infty} \Psi_i\} = \{\ell : \ell \in \Psi_n\}$ for some smallest number $n \in \mathbb{N}_0$. The induction proof ensures the existence of some (S_n, E_n) such that $\Psi_n \subseteq Th(S_n)$. Because $\{\ell : \ell \in \Psi_n\} = T$ is a state, this implies $T = S_n$. Consequently, $((S \setminus C) \cup E, E) \overset{*}{\leadsto}_{\mathcal{R}} (T, E_n)$, that is, T is causal successor state. *Qed.*

While this result shows that causal relationships allow to obtain any causal minimizing-change successor, the converse is not true. That is, there may exist causal successor states which do not obey the request for minimizing change. To illustrate this, we further extend (for the last time) our electric circuit. This new example shows that minimizing change might be a policy too restrictive to account for all possible successor states. It is obviously necessary to consider a non-deterministic action to this end, for there must be at least two successor states, one of which is non-minimal.

Example 2.6.3. The electric circuit of Example 2.3.2, depicted in Fig. 2.4, is slightly modified and further augmented by a light detecting device that becomes and stays activated as soon as light turns on; see Fig. 2.8. Accordingly, the basic action domain of Example 2.3.2 is extended by fluent name $\texttt{detect}^0$. The new arrangement is formalized by these four state constraints:

$$\begin{array}{l} \texttt{light} \equiv \texttt{up}(\texttt{s}_1) \wedge \texttt{up}(\texttt{s}_2) \\ \texttt{relay} \equiv \texttt{up}(\texttt{s}_1) \wedge \texttt{up}(\texttt{s}_3) \\ \texttt{relay} \supset \neg\texttt{up}(\texttt{s}_2) \\ \texttt{light} \supset \texttt{detect} \end{array} \tag{2.13}$$

Obviously, potential influence is as in our predecessor circuit but with additionally having that the state of the light bulb may affect the detector. After enhancing influence information $\mathcal{I}$ of Example 2.3.2 by pair (light, detect), the above state constraints determine the following causal relationships $\mathcal{R}$ according to our generation procedure.

$$\begin{array}{ll} \texttt{up}(\texttt{s}_1)\ \underline{\text{causes}}\ \texttt{light}\ \underline{\text{if}}\ \texttt{up}(\texttt{s}_2) & \texttt{up}(\texttt{s}_2)\ \underline{\text{causes}}\ \texttt{light}\ \underline{\text{if}}\ \texttt{up}(\texttt{s}_1) \\ \neg\texttt{up}(\texttt{s}_1)\ \underline{\text{causes}}\ \neg\texttt{light}\ \underline{\text{if}}\ \top & \neg\texttt{up}(\texttt{s}_2)\ \underline{\text{causes}}\ \neg\texttt{light}\ \underline{\text{if}}\ \top \end{array}$$

$$\begin{array}{ll} \texttt{up}(\texttt{s}_1)\ \underline{\text{causes}}\ \texttt{relay}\ \underline{\text{if}}\ \texttt{up}(\texttt{s}_3) & \texttt{up}(\texttt{s}_3)\ \underline{\text{causes}}\ \texttt{relay}\ \underline{\text{if}}\ \texttt{up}(\texttt{s}_1) \\ \neg\texttt{up}(\texttt{s}_1)\ \underline{\text{causes}}\ \neg\texttt{relay}\ \underline{\text{if}}\ \top & \neg\texttt{up}(\texttt{s}_3)\ \underline{\text{causes}}\ \neg\texttt{relay}\ \underline{\text{if}}\ \top \end{array}$$

$$\begin{array}{ll} \texttt{relay}\ \underline{\text{causes}}\ \neg\texttt{up}(\texttt{s}_2)\ \underline{\text{if}}\ \top & \texttt{light}\ \underline{\text{causes}}\ \texttt{detect}\ \underline{\text{if}}\ \top \end{array}$$

Now, suppose we toggle the first switch, $\texttt{s}_1$, in the state depicted in Fig. 2.8. What is the expected outcome? Obviously, the relay gets activated and, then,

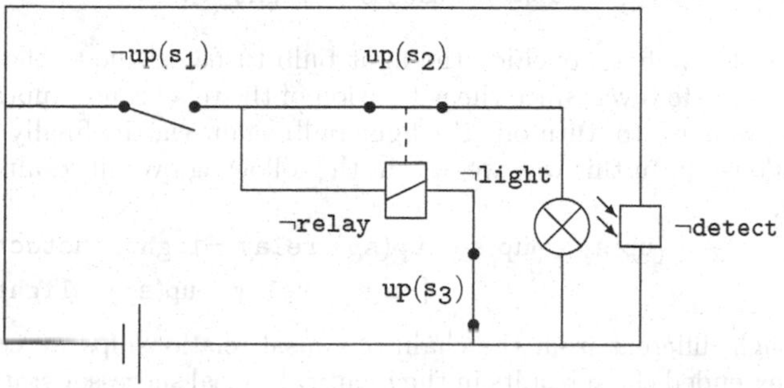

Figure 2.8. A modified electric circuit (c.f. Fig. 2.4) augmented by a device, represented by fluent **detect**, which registers an activation of the light bulb (this device combines a phototransistor and flipflop). It is assumed that no action of light has occurred yet.

attracts the second switch, $\mathtt{sw}_2$. Hence, the latter is open in the finally resulting state. Notice, however, that as soon as the first switch is in the upper position, the sub-circuit involving the light bulb gets closed. This may activate the light bulb for an instant, that is, before the second switch jumps its position as a result of the relay activation. If this is indeed the case, then this 'flash' might be registered by the photo device. Hence, while it is clear that the light bulb is off in the resulting state, fluent **detect** possibly becomes true.[17] Accordingly, two causal successor states exist for initial state $\{\neg\mathtt{up}(\mathtt{s}_1), \mathtt{up}(\mathtt{s}_2), \mathtt{up}(\mathtt{s}_3), \neg\mathtt{light}, \neg\mathtt{relay}, \neg\mathtt{detect}\}$ and action $\mathtt{toggle}(\mathtt{s}_1)$, which are obtained as follows. The unique preliminary successor state along with the corresponding direct effect is

$$(\{\mathtt{up}(\mathtt{s}_1), \mathtt{up}(\mathtt{s}_2), \mathtt{up}(\mathtt{s}_3), \neg\mathtt{light}, \neg\mathtt{relay}, \neg\mathtt{detect}\}, \{\mathtt{up}(\mathtt{s}_1)\}) \qquad (2.14)$$

The first component violates both the first and the second constraint of (2.13). There are several ways to proceed. First, we can apply causal relationship $\mathtt{up}(\mathtt{s}_1)$ <u>causes</u> $\mathtt{relay}$ <u>if</u> $\mathtt{up}(\mathtt{s}_3)$ followed by $\mathtt{relay}$ <u>causes</u> $\neg\mathtt{up}(\mathtt{s}_2)$ <u>if</u> $\top$, which results in

$$\begin{aligned}(\{\mathtt{up}(\mathtt{s}_1), \neg\mathtt{up}(\mathtt{s}_2), \mathtt{up}(\mathtt{s}_3), \mathtt{relay}, \neg\mathtt{light}, \neg\mathtt{detect}\}\,, \\ \{\mathtt{up}(\mathtt{s}_1), \mathtt{relay}, \neg\mathtt{up}(\mathtt{s}_2)\}\,)\end{aligned}$$

The first component satisfies all of the state constraints and, consequently, is a causal successor state.

Another possibility to process the state-effect pair (2.14) is to apply the following sequence of causal relationships:

$$\begin{array}{rlll} \mathtt{up}(\mathtt{s}_1) & \underline{\text{causes}} & \mathtt{light} & \underline{\text{if}}\ \mathtt{up}(\mathtt{s}_2) \\ \mathtt{up}(\mathtt{s}_1) & \underline{\text{causes}} & \mathtt{relay} & \underline{\text{if}}\ \mathtt{up}(\mathtt{s}_3) \\ \mathtt{relay} & \underline{\text{causes}} & \neg\mathtt{up}(\mathtt{s}_2) & \underline{\text{if}}\ \top \\ \neg\mathtt{up}(\mathtt{s}_2) & \underline{\text{causes}} & \neg\mathtt{light} & \underline{\text{if}}\ \top \end{array} \qquad (2.15)$$

In words, we first conclude the light bulb turns on due to the second switch being on. However, since the activation of the relay causes $\mathtt{up}(\mathtt{s}_2)$ to become false, we have to 'turn off' the light bulb again via the finally applied causal relationship. In this way we obtain the following overall resulting state-effect pair.

$$\begin{aligned}(\{\mathtt{up}(\mathtt{s}_1), \neg\mathtt{up}(\mathtt{s}_2), \mathtt{up}(\mathtt{s}_3), \mathtt{relay}, \neg\mathtt{light}, \neg\mathtt{detect}\}\,, \\ \{\mathtt{up}(\mathtt{s}_1), \mathtt{relay}, \neg\mathtt{up}(\mathtt{s}_2), \neg\mathtt{light}\}\,)\end{aligned}$$

Though different from the chain of causal relationships we considered first, the extended chain results in the identical causal successor state. But we have

[17] What actually happens depends on the physical properties of both the relay and the light detector. Lacking this precise information, common sense interprets the situation as giving rise to non-determinism. See the following section for a detailed discussion on this phenomenon.

not considered all possibilities yet. Notice that light becomes true and is among the current effects after having applied the first one of the causal relationships (2.15), and it stays true until the last one is applied. Therefore we may insert causal relationship light causes detect if $\top$ somewhere in between. This additionally causes detect to become true, that is, the following state-effect pair can also be generated:

$$(\{\texttt{up(s}_1\texttt{)}, \neg\texttt{up(s}_2\texttt{)}, \texttt{up(s}_3\texttt{)}, \texttt{relay}, \neg\texttt{light}, \texttt{detect}\}\ , \\ \{\texttt{up(s}_1\texttt{)}, \texttt{detect}, \texttt{relay}, \neg\texttt{up(s}_2\texttt{)}, \neg\texttt{light}\}\ )$$

Its first component is acceptable and different from the causal successor above. No further causal successor states can be obtained.

Now, the first one of the two successors, i.e.,

$$T_1 = \{\texttt{up(s}_1\texttt{)}, \neg\texttt{up(s}_2\texttt{)}, \texttt{up(s}_3\texttt{)}, \texttt{relay}, \neg\texttt{light}, \neg\texttt{detect}\}$$

is strictly closer to initial state S than the second one, i.e.,

$$T_2 = \{\texttt{up(s}_1\texttt{)}, \neg\texttt{up(s}_2\texttt{)}, \texttt{up(s}_3\texttt{)}, \texttt{relay}, \neg\texttt{light}, \texttt{detect}\}$$

This indicates that the latter cannot be a causal minimizing-change successor. Indeed, while T_1 is fixpoint of $\lambda T. \{\ell : ((S \cap T) \cup E, E) \Vdash_{\mathcal{R}} \ell\}$ (with $E = \{\texttt{up(s}_1\texttt{)}\}$), state T_2 is not because

$$((S \cap T_2) \cup E, E) = (\{\texttt{up(s}_3\texttt{)}, \neg\texttt{light}, \texttt{up(s}_1\texttt{)}\}, \{\texttt{up(s}_1\texttt{)}\} \nVdash_{\mathcal{R}} \texttt{detect}$$

■

This domain proves that minimization is not always a concept adequate for distinguishing between possible indirect effects on the one hand, and unfounded changes on the other hand. This observation challenges the common belief that minimizing change is essential for solving the Ramification Problem. In fact, the aim of generating ramifications is not to minimize change but to avoid changes that are not caused, which, as we have seen, need not be identical goals.

2.7 Triggered vs. Coupled Effects

The instructive example concluding the preceding section shows how causal relationships often literally run a race. A different finish sometimes means a different successor state, thus giving rise to uncertainty in form of nondeterminism. This raises the question whether it might be overly credulous to consider possible any order in which these relationships are applied. Not all computed chains of indirect effects may be equally likely to happen in reality. This would be the case if the causal lag between some particular indirect effect and its triggering cause is generally shorter than the causal lag between another particular effect and its cause. Suppose, for instance, our relay is known to take its time for warming up until it reaches the power to attract the switch, whereas the light detector registers the shortest flash with utmost alertness. Then it seems unreasonable to consider the possibility that the detector stays off when toggling switch $\mathtt{s}_1$ in the situation depicted in Fig. 2.8.

One approach to this problem is to introduce an explicit notion of time, namely, in specifying the exact delay between the occurrence of an effect and its cause. This, however, is not in the spirit of the Ramification Problem. For the latter is concerned with accounting for those indirect effects which so rapidly follow the performance of an action that common sense considers them virtually instantaneous. Had we assumed precise knowledge of the relay's delay, then its attracting switches fell into the category of so-called "delayed" effects. Though certainly of importance, too, these effects need to be meticulously distinguished from those indirect effects investigated in the context of the Ramification Problem. For delayed effects deserve a separate state transition, that is, they should not occur during the same transition step as its triggering effect. In contrast, solving the Ramification Problem means to account for all indirect effects that are summarized in a single state transition. This amounts to performing qualitative reasoning about causal lags, as opposed to quantitative reasoning, which would require precise knowledge of virtually indistinguishable time intervals. Qualitative reasoning, which acknowledges the fact that common sense often lacks precise knowledge, considers equal all causal lags greater than zero.

The very last remark, however, reveals one critical aspect which we have neglected hitherto. Causal lags may indeed differ even qualitatively, namely, in possibly being zero. An important category of indirect effects obviously of this nature is effects triggered by so-called definitional state constraints, which we have already touched upon in Section 2.4. Recall constraint $\forall x[\mathtt{down}(x) \equiv \neg\mathtt{up}(x)]$. Having some instance like $\mathtt{up}(\mathtt{s}_1)$ as direct or indirect effect, this gives rise to additional indirect effect $\neg\mathtt{down}(\mathtt{s}_1)$. The causal lag between these two effects is zero—not even for the tiniest fraction of time a state is imaginable where $\mathtt{up}(\mathtt{s}_1)$ already and $\mathtt{down}(\mathtt{s}_1)$ still hold. The total absence of a causal lag qualitatively distinguishes this ramification from most ones we have considered throughout the previous sections. Let us call *coupled* all effects which occur with causal lag zero, as opposed to *triggered* effects. Failing to account for this distinction may lead to surprising conclusions even if granted that, as we have argued, temporal differences between causal lags are to be neglected. The following scenario illustrates why coupled and triggered effects must not be treated alike.

Example 2.7.1. Suppose a bowl well filled with soup is standing on a rectangular table; see Fig. 2.9. Whenever the left hand side of the table is lifted up but not the right hand side (or vice versa), then the spilling soup stains the tablecloth. If, however, both sides are lifted up simultaneously, then no soup is expected to spill out. Let us model this scenario by a basic action domain consisting of entities `lhs` and `rhs` (left and right hand side of the table),

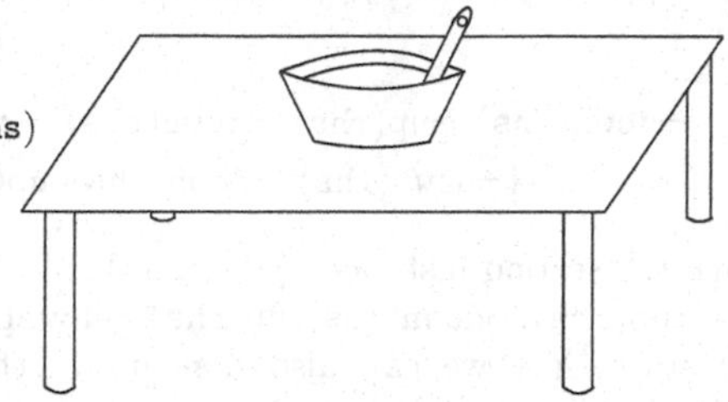

Figure 2.9. A bowl filled with soup is standing on a table. The soup spills out and produces a stain if the table is lifted on one hand side but not the other. Nothing of this sort is expected when lifting up the table on both sides simultaneously.

fluent names $\texttt{up}^1$, $\texttt{down}^1$, and $\texttt{stain}^0$, action name $\texttt{lift-both-sides}^0$, and action law

$$\begin{aligned}\texttt{lift-both-sides}\ \underline{\text{transforms}}\ &\{\texttt{down(lhs)}, \texttt{down(rhs)}\}\\ \underline{\text{into}}\ &\{\neg\texttt{down(lhs)}, \neg\texttt{down(rhs)}\}\end{aligned}$$

Furthermore, let the set of state constraints consist of

$$\begin{aligned}&\forall x\, [\,\texttt{down}(x) \equiv \neg\texttt{up}(x)\,]\\ &\texttt{up(lhs)} \wedge \neg\texttt{up(rhs)} \vee \texttt{up(rhs)} \wedge \neg\texttt{up(lhs)} \supset \texttt{stain}\end{aligned}$$

Influence information $\mathcal{I} = \{(\texttt{up}(x), \texttt{down}(x)), (\texttt{down}(x), \texttt{up}(x)), (\texttt{up}(x), \texttt{stain})\}$ plus the constraints determine the following causal relationships:

$$\begin{array}{rl}\texttt{down}(x)\ \underline{\text{causes}}\ \neg\texttt{up}(x)\ \ \underline{\text{if}}\ \top & \texttt{up(lhs)}\ \underline{\text{causes}}\ \texttt{stain}\ \underline{\text{if}}\ \neg\texttt{up(rhs)}\\ \texttt{up}(x)\ \underline{\text{causes}}\ \neg\texttt{down}(x)\ \underline{\text{if}}\ \top & \texttt{up(rhs)}\ \underline{\text{causes}}\ \texttt{stain}\ \underline{\text{if}}\ \neg\texttt{up(lhs)}\\ \neg\texttt{down}(x)\ \underline{\text{causes}}\ \texttt{up}(x)\ \ \underline{\text{if}}\ \top & \neg\texttt{up(rhs)}\ \underline{\text{causes}}\ \texttt{stain}\ \underline{\text{if}}\ \texttt{up(lhs)}\\ \neg\texttt{up}(x)\ \underline{\text{causes}}\ \texttt{down}(x)\ \ \underline{\text{if}}\ \top & \neg\texttt{up(lhs)}\ \underline{\text{causes}}\ \texttt{stain}\ \underline{\text{if}}\ \texttt{up(rhs)}\end{array}$$

(The interested reader may verify this being the output of our algorithm of Fig. 2.6.)

Now, consider the state where the table stands firmly on the floor and there is no stain, $S = \{\neg\texttt{up(lhs)}, \texttt{down(lhs)}, \neg\texttt{up(rhs)}, \texttt{down(rhs)}, \neg\texttt{stain}\}$, as depicted in Fig. 2.9. Performing `lift-both-sides` results in the preliminary successor

$$S' = \{\neg\texttt{up(lhs)}, \neg\texttt{down(lhs)}, \neg\texttt{up(rhs)}, \neg\texttt{down(rhs)}, \neg\texttt{stain}\}$$

obtained through the direct effect $E = \{\neg\texttt{down(lhs)}, \neg\texttt{down(rhs)}\}$. This preliminary state violates the state constraints. The two possible instances of causal relationship $\neg\mathtt{down}(x)$ causes $\mathtt{up}(x)$ if $\top$ are applicable to the state-effect pair (S', E). Suppose we first apply instance $\{x \mapsto \mathtt{lhs}\}$. This yields the state-effect pair

$$(\{\mathtt{up(lhs)}, \neg\mathtt{down(lhs)}, \neg\mathtt{up(rhs)}, \neg\mathtt{down(rhs)}, \neg\mathtt{stain}\} ,\\ \{\neg\mathtt{down(lhs)}, \neg\mathtt{down(rhs)}, \mathtt{up(lhs)}\})$$

We can proceed with the second instance, $\{x \mapsto \mathtt{rhs}\}$, of the above relationship, thus obtaining $\{\mathtt{up(lhs)}, \neg\mathtt{down(lhs)}, \mathtt{up(rhs)}, \neg\mathtt{down(rhs)}, \neg\mathtt{stain}\}$ as a causal successor state. Yet we can also first insert the application of $\mathtt{up(lhs)}$ causes $\mathtt{stain}$ if $\neg\mathtt{up(rhs)}$. Followed by the conversion of $\neg\mathtt{up(rhs)}$, we thus obtain $\{\mathtt{up(lhs)}, \neg\mathtt{down(lhs)}, \mathtt{up(rhs)}, \neg\mathtt{down(rhs)}, \mathtt{stain}\}$ as another possible causal successor, where surprisingly a stain has been produced! The identical two causal successor states can be obtained in case relationship instance $\neg\mathtt{down(rhs)}$ causes $\mathtt{up(rhs)}$ if $\top$ is applied first, allowing to produce a stain via $\mathtt{up(rhs)}$ causes $\mathtt{stain}$ if $\neg\mathtt{up(lhs)}$. ∎

The general problem which causes the unexpected possibility of a stain here, is that causal relationships introduce an artificial causal lag between coupled effects. Some other causal relationship may exploit this lag for 'squeezing in' a triggered effect which can never occur in reality. As a consequence, our approach to the Ramification Problem needs to be refined in that it respects the distinction between virtually and truly instantaneous indirect effects.

To this end, some state constraints are distinguished as being *steady*. All definitional constraints enjoy this property, but others might as well. Consider, as an example, the situation depicted in Fig. 2.10. The corresponding state constraint $\forall x\, [\, \mathtt{location}(\mathtt{a}, x) \equiv \mathtt{location}(\mathtt{b}, x)\,]$ is to be regarded as steady. The criterion for characterizing a state constraint as steady is that not even for an instant a situation is imaginable where this constraint is violated. All indirect effects determined by steady state constraints have causal lag zero, i.e., are coupled. As argued, during the application of causal relationships the insertion of an effect with real causal lag in between the generation of coupled effects needs to be prohibited. This is achieved by first applying only causal relationships stemming from steady state constraints, until none of these constraints is violated. Only thereafter a triggered effect may be generated, again followed by accounting for all coupled effects necessary to satisfy the steady constraints, and so on until an overall acceptable state obtains. This strategy is formalized in the following definition of successor state.

Definition 2.7.1. *Let $\mathcal{E}$, $\mathcal{F}$, $\mathcal{A}$, and $\mathcal{L}$ be sets of entities, fluent names, action names, and action laws, respectively. Furthermore, let $\mathcal{C}$ be a set of state constraints and $\mathcal{R}$ a set of causal relationships, and let $\mathcal{C}_s \subseteq \mathcal{C}$ and $\mathcal{R}_s \subseteq \mathcal{R}$ be designated sets of steady constraints and relationships, respectively. If S is an acceptable state and a an action, then a state S' is a* successor *of S*

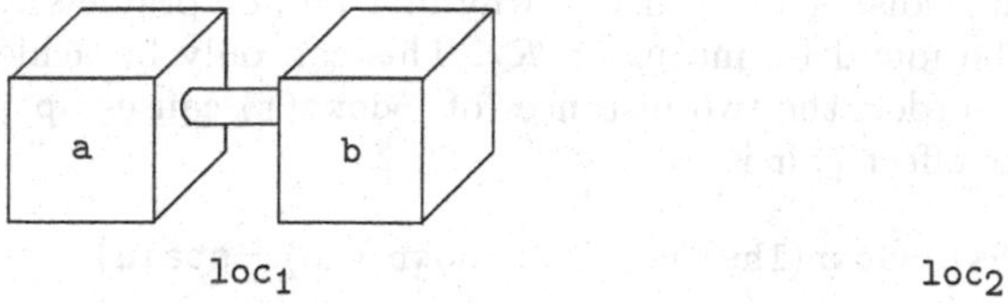

Figure 2.10. These two blocks are connected by a firm bar. Whenever one of them changes its location, e.g., from $\texttt{loc}_1$ to $\texttt{loc}_2$, then the other block moves as well. Not even for an instant can the two blocks occupy different locations. Hence the state constraint $\forall x\,[\,\texttt{location}(\texttt{a},x) \equiv \texttt{location}(\texttt{b},x)\,]$ is to be considered steady.

and a iff the following holds: Set $\mathcal{L}$ contains an applicable action law instance a transforms C into E and there exist states $S_0, S_0', \ldots, S_n, S_n'$ and sets of fluent literals $E_0, E_0', \ldots, E_n, E_n'$ $(n \geq 0)$ such that $S_0 = (S \setminus C) \cup E$, $E_0 = E$,

$$(S_0, E_0) \overset{*}{\leadsto}_{\mathcal{R}_s} (S_0', E_0') \leadsto_{\mathcal{R}} (S_1, E_1) \overset{*}{\leadsto}_{\mathcal{R}_s} (S_1', E_1') \leadsto_{\mathcal{R}} \cdots \overset{*}{\leadsto}_{\mathcal{R}_s} (S_n', E_n')$$

and, for each $0 \leq i \leq n$, S_i' is acceptable wrt. $\mathcal{C}_s$ and S_n' is acceptable. ■

Example 2.7.2. Consider $\mathcal{E} = \{\texttt{lhs}, \texttt{rhs}\}$, $\mathcal{F} = \{\texttt{up}^1, \texttt{down}^1, \texttt{stain}^0\}$, and $\mathcal{A} = \{\texttt{lift-both-sides}^0\}$, and let $\mathcal{L}$ consist of

$$\begin{array}{lll} \texttt{lift-both-sides} & \underline{\text{transforms}} & \{\texttt{down(lhs)}, \texttt{down(rhs)}\} \\ & \underline{\text{into}} & \{\neg\texttt{down(lhs)}, \neg\texttt{down(rhs)}\} \end{array}$$

as above. Furthermore, let $\mathcal{C}_s = \{\forall x\,[\,\texttt{down}(x) \equiv \neg\texttt{up}(x)\,]\}$, and let $\mathcal{C}$ be $\mathcal{C}_s$ plus $\texttt{up(lhs)} \land \neg\texttt{up(rhs)} \lor \texttt{up(rhs)} \land \neg\texttt{up(lhs)} \supset \texttt{stain}$. Then the steady causal relationships $\mathcal{R}_s$ are

$$\begin{array}{ll} \texttt{down}(x)\ \underline{\text{causes}}\ \neg\texttt{up}(x)\ \underline{\text{if}}\ \top & \texttt{up}(x)\ \underline{\text{causes}}\ \neg\texttt{down}(x)\ \underline{\text{if}}\ \top \\ \neg\texttt{down}(x)\ \underline{\text{causes}}\ \texttt{up}(x)\ \underline{\text{if}}\ \top & \neg\texttt{up}(x)\ \underline{\text{causes}}\ \texttt{down}(x)\ \underline{\text{if}}\ \top \end{array}$$

and $\mathcal{R}$ is $\mathcal{R}_s$ plus

$$\begin{array}{ll} \texttt{up(lhs)}\ \underline{\text{causes}}\ \texttt{stain}\ \underline{\text{if}}\ \neg\texttt{up(rhs)} & \neg\texttt{up(rhs)}\ \underline{\text{causes}}\ \texttt{stain}\ \underline{\text{if}}\ \texttt{up(lhs)} \\ \texttt{up(rhs)}\ \underline{\text{causes}}\ \texttt{stain}\ \underline{\text{if}}\ \neg\texttt{up(lhs)} & \neg\texttt{up(lhs)}\ \underline{\text{causes}}\ \texttt{stain}\ \underline{\text{if}}\ \texttt{up(rhs)} \end{array}$$

As above, suppose $S = \{\neg\texttt{up(lhs)}, \texttt{down(lhs)}, \neg\texttt{up(rhs)}, \texttt{down(rhs)}, \neg\texttt{stain}\}$ be the current state. Performing lift-both-sides produces the following state-effect pair.

$$(\ \{\neg\texttt{up(lhs)}, \neg\texttt{down(lhs)}, \neg\texttt{up(rhs)}, \neg\texttt{down(rhs)}, \neg\texttt{stain}\}\ ,\ \{\neg\texttt{down(lhs)}, \neg\texttt{down(rhs)}\}\ )$$

Its first component violates $\mathcal{C}_s$, which is why first an acceptable state with this regard is to be found by means of $\mathcal{R}_s$. This can only be achieved by applying, in either order, the two instances of $\neg\mathtt{down}(x)$ causes $\mathtt{up}(x)$ if $\top$. The resulting state-effect pair is

$$(\{\mathtt{up(lhs)}, \neg\mathtt{down(lhs)}, \mathtt{up(rhs)}, \neg\mathtt{down(rhs)}, \neg\mathtt{stain}\} , \\ \{\neg\mathtt{down(lhs)}, \neg\mathtt{down(rhs)}, \mathtt{up(lhs)}, \mathtt{up(rhs)}\})$$

No further causal relationship, steady or non-steady, is applicable. The first component satisfies the entire set of state constraints $\mathcal{C}$, hence constitutes the unique successor state. ■

Since now we finally arrived at a satisfactory solution to the Ramification Problem, let us extend the formal concepts of action domains, scenarios, and entailment so as to be able to specify and reason about domains involving indirect effects of actions.

Definition 2.7.2. *A* ramification domain $\mathcal{D}$ *is a 6-tuple* $(\mathcal{E}, \mathcal{F}, \mathcal{A}, \mathcal{L}, \mathcal{C}, \mathcal{R})$ *where* $(\mathcal{E}, \mathcal{F}, \mathcal{A}, \mathcal{L})$ *constitutes a basic action domain,* $\mathcal{C}$ *is a set of state constraints with a designated subset* $\mathcal{C}_s$ *of steady constraints, and* $\mathcal{R}$ *is a set of causal relationships with a designated subset* $\mathcal{R}_s$ *of steady relationships. The* transition model *of* $\mathcal{D}$ *is a mapping* Σ *from pairs of an acceptable state and an action into (possibly empty) sets of states such that* $S' \in \Sigma(S, a)$ *iff* S' *is successor of* S *and* a. ■

As solutions to the Ramification Problem are only concerned with the sophistication of transition models, all concepts beyond can be adopted without modification. For the sake of completeness, let us reiterate them.

Definition 2.7.3. *A* ramification scenario *is a pair* $(\mathcal{O}, \mathcal{D})$ *where* $\mathcal{D}$ *is a ramification domain and* $\mathcal{O}$ *is a set of observations. An* interpretation *for* $(\mathcal{O}, \mathcal{D})$ *is a pair* (Σ, Res) *where* Σ *is the transition model of* $\mathcal{D}$ *and* Res *is a partial function which maps finite (possibly empty) action sequences to acceptable states and which satisfies the following:*

1. $Res([\,])$ *is defined.*
2. *For any sequence* $a^* = [a_1, \ldots, a_{k-1}, a_k]$ *of actions* $(k > 0)$,
   a) $Res(a^*)$ *is defined if and only if* $Res([a_1, \ldots, a_{k-1}])$ *is defined and* $\Sigma(Res([a_1, \ldots, a_{k-1}]), a_k)$ *is not empty, and*
   b) $Res(a^*) \in \Sigma(Res([a_1, \ldots, a_{k-1}]), a_k)$.

A model *of a ramification scenario* $(\mathcal{O}, \mathcal{D})$ *is an interpretation* (Σ, Res) *such that each* $o \in \mathcal{O}$ *is true in* Res*, and an observation* o *is* entailed, *written* $\mathcal{O} \models_{\mathcal{D}} o$*, iff it is true in all models.* ■

As a small exercise, the reader may consider the ramification domain modeling the electric circuit with the relay of Fig. 2.4, along with the scenario given by these two observations:

$$\begin{array}{ll} \texttt{up}(\texttt{s}_2) & \underline{\text{after}}\ [\,] \\ \neg\texttt{relay} & \underline{\text{after}}\ [\texttt{toggle}(\texttt{s}_3)] \end{array}$$

What can be concluded as to the initial state of the light bulb here? That is, is either one of the two observations **light** after [] or ¬**light** after [] entailed?[18]

2.8 Implicit Qualifications vs. Indirect Effects

Sometimes the available causal relationships fail to produce even a single acceptable state. Rather than indicating an erroneous specification, the non-existence of a successor state tells us something about the executability of the action at hand. Recall that if no preliminary successor state exists, that is, if no action law is applicable, then this is an indication that some action precondition is not met. This interpretation transfers to the case where a preliminary successor exists but no overall successor state. Again, the action at hand is, for some reason, not qualified. That is, aside from the preconditions explicitly mentioned in action laws, additional, *implicit* preconditions are being revealed by the failure to generate an acceptable state. In causing the unsatisfiability of the state constraints, these supplementary action qualifications can be said to derive from state constraints—just like ramifications do. Although implicit preconditions are somehow opposite to indirect effects, being able to encounter them is vital when solving the Ramification Problem. For if a successor state is suggested in case some implicit qualification is not met, then this state is necessarily based on indirect effects which could never occur in reality.

An exemplary scenario shall emphasize the distinction between implicit qualifications and indirect effects. An interesting feature here is that one and the same state constraint gives rise to both qualification and ramification.

Example 2.8.1. The following is an extended version of the famous *Yale Shooting* domain, which describes a turkey-hunting. Suppose given the only entity **turkey**, the two fluent names $\texttt{alive}^1$ and $\texttt{fleeing}^1$, and the two action names $\texttt{shoot}^1$ and $\texttt{startle}^1$, which are accompanied by these two action laws:

$$\begin{array}{llll} \texttt{shoot}(x) & \underline{\text{transforms}} & \{\texttt{alive}(x)\} & \underline{\text{into}}\ \{\neg\texttt{alive}(x)\} \\ \texttt{startle}(x) & \underline{\text{transforms}} & \{\neg\texttt{fleeing}(x)\} & \underline{\text{into}}\ \{\texttt{fleeing}(x)\} \end{array}$$

[18] The right answer is that **light** after [] is entailed: Let (Σ, Res) be any model of the scenario. The two state constraints $\texttt{relay} \equiv \neg\texttt{up}(\texttt{s}_1) \wedge \texttt{up}(\texttt{s}_3)$ and $\texttt{relay} \supset \neg\texttt{up}(\texttt{s}_2)$ in conjunction with $\texttt{up}(\texttt{s}_2) \in Res([\,])$ imply that $\texttt{up}(\texttt{s}_1) \vee \neg\texttt{up}(\texttt{s}_3) \in Res([\,])$. Now, had we $\neg\texttt{up}(\texttt{s}_1) \in Res([\,])$ it would follow that $\neg\texttt{up}(\texttt{s}_3) \in Res([\,])$ and, hence, $\texttt{up}(\texttt{s}_3), \neg\texttt{up}(\texttt{s}_1) \in Res([\texttt{toggle}(\texttt{s}_3)])$, which contradicts $\neg\texttt{relay} \in Res([\texttt{toggle}(\texttt{s}_3)])$. Thus we find that $\texttt{up}(\texttt{s}_1) \in Res([\,])$, which, given $\texttt{up}(\texttt{s}_2) \in Res([\,])$, implies $\texttt{light} \in Res([\,])$ according to state constraint $\texttt{light} \equiv \texttt{up}(\texttt{s}_1) \wedge \texttt{up}(\texttt{s}_2)$.

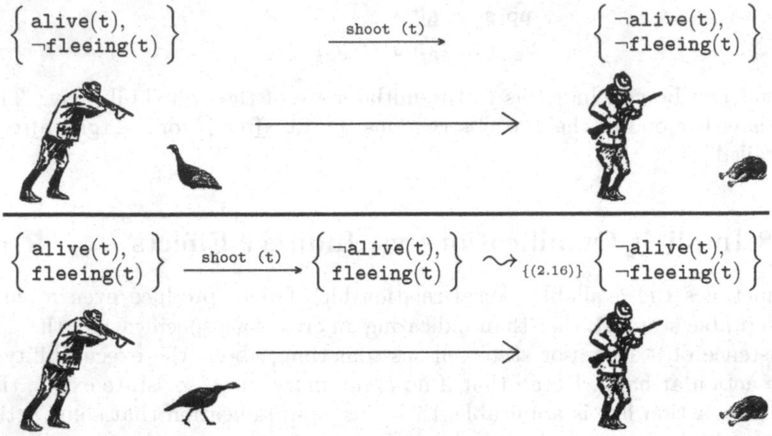

Figure 2.11. Shooting the turkey (t) always has the direct effect ¬alive(t). If the turkey is in motion, then the action additionally causes ¬fleeing(t).

The (steady) state constraint $\forall x[\texttt{fleeing}(x) \supset \texttt{alive}(x)]$ restricts fleeing subjects to vivid ones. The correct information of influence in this context is $\mathcal{I} = \{(\texttt{alive}(x), \texttt{fleeing}(x))\}$, that is, a change of $\texttt{alive}(x)$ might affect the truth-value of $\texttt{fleeing}(x)$ but not vice versa. The constraint thus determines a single causal relationship, viz.

$$\neg\texttt{alive}(x) \ \underline{\text{causes}} \ \neg\texttt{fleeing}(x) \ \underline{\text{if}} \ \top \qquad (2.16)$$

Suppose action shoot(turkey) is performed in a state where the victim is still alive. In case the initial state is $\{\texttt{alive(turkey)}, \neg\texttt{fleeing(turkey)}\}$, it suffices to generate the direct effect, $\{\neg\texttt{alive(turkey)}\}$, of the corresponding action law instance. If, on the other hand, the initial state is $\{\texttt{alive(turkey)}, \texttt{fleeing(turkey)}\}$, then the available causal relationship needs to be applied to accommodate an additional effect: Not only does the turkey drop dead, it also stops fleeing. The two situations are illustrated in Fig. 2.11. We see that our state constraint may give rise to indirect effects.

In contrast, suppose we perform action startle(turkey) in a situation where the turkey idles. As above, if $\{\texttt{alive(turkey)}, \neg\texttt{fleeing(turkey)}\}$ is the initial state, then the direct effect, fleeing(turkey), of the corresponding action law instance suffices to obtain an acceptable state. On the other hand, consider initial state $S = \{\neg\texttt{alive(turkey)}, \neg\texttt{fleeing(turkey)}\}$. The only preliminary successor of S and action startle(turkey) is state $S' = \{\neg\texttt{alive(turkey)}, \texttt{fleeing(turkey)}\}$, obtained through the direct effect $E = \{\texttt{fleeing(turkey)}\}$. State S' violates our constraint. Moreover, the only available causal relationship is not applicable here since $\neg\texttt{alive(turkey)} \notin E$. Consequently, no successor state exists. In other words, our state constraint enforces the additional precondition that the turkey must

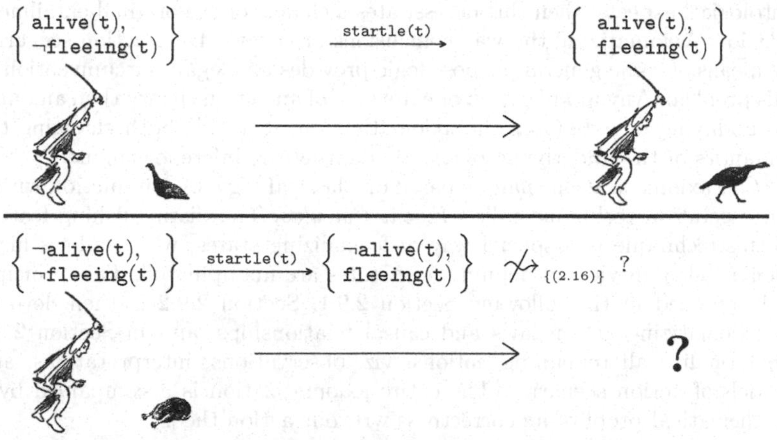

Figure 2.12. Startling the turkey is possible only if the animal is alive. This implicit precondition is encountered by failing to transform the preliminary successor {¬alive(t), fleeing(t)} into an acceptable state via the only available causal relationship (2.16).

be alive if we want to startle it with success; see Fig. 2.12. We see that our constraint may also give rise to implicit qualifications. ■

This completes our formal account of the Ramification Problem. Let us summarize: A ramification domain contains a number of state constraints, of which some are steady, that give rise to indirect effects of actions. With the aid of general information as to possible influences among the fluents, the potential indirect effects, formalized as so-called causal relationships, can be automatically extracted from the constraints. This helps to distinguish the correct effects from both unmotivated changes and implicit qualifications. Causal relationships are successively applied to preliminary successor states until all indirect effects have been accounted for; in the course of this process the so-called coupled effects, deriving from steady state constraints, are to be generated prior to any so-called triggered effect.

2.9 A Fluent Calculus Axiomatization

In this section, we develop an axiomatization of our action theory by means of standard logic. The major motivation for so doing is that action theories, as they stand, are less suited than general purpose logics when it comes to automating reasoning. The reason is that entailment relations of action theories are usually much more specialized and sophisticated because they reflect complex notions such as time, change, and causality. Moreover, whenever an action theory is modified or extended in order to address additional ontological aspects, then this necessitates a change of the original entailment relation. Axiomatizing the way conclusions are drawn from action scenarios by means of some general purpose logic provides an elegant circumvention of this problem. Any modification or extension of an action theory then amounts to modifying or extending the axiomatization; it avoids both changing the semantics of the underlying logic and adapting the inference engine.

Our axiomatization shall be based on classical logic in conjunction with a representation technique called *Fluent Calculus*. The distinguishing feature of this technique is a special way of formalizing states, which allows highest flexibility in view of defining how states are manipulated. This principle is introduced in the following Section 2.9.1. Section 2.9.2 is then devoted to axiomatizing action laws and causal relationships, and in Section 2.9.3 we formalize all remaining notions, viz. observations, interpretations, and models of action scenarios. The entire axiomatization is accompanied by a mathematical proof of its correctness wrt. our action theory.

2.9.1 Reifying States

The common principle of Fluent Calculus-based formalizations of action theories is a particular axiomatization of the fundamental entity, namely, the state. A state is encoded as a term, accordingly called *state term*, which contains as sub-terms the fluent literals that are true in that state. This enables the use of first-order formulas to define how state terms evolve in the course of time. Representing fluents by terms instead of predicates follows a standard technique in logic known as *reification*, which generally allows for a restricted form of meta-reasoning without appealing to higher-order logic nor to extensions of classical logic. Sub-terms representing fluent literals compose state terms by employing a special binary function, which is illustratively denoted by the symbol "$\circ$" and is written in infix notation. For example, a term representation of the state $\{\neg\mathtt{up}(\mathtt{s_1}), \mathtt{up}(\mathtt{s_2}), \neg\mathtt{light}\}$ is

$$(\neg\mathtt{up}(\mathtt{s_1}) \circ \mathtt{up}(\mathtt{s_2})) \circ \neg\mathtt{light} \qquad (2.17)$$

where formally each n-place fluent becomes an n-place function, each entity becomes a constant, and where the negation symbol is a distinguished unary function.

Obviously, it should be irrelevant at which position a fluent literal occurs in a state term. That is, (2.17) and $\mathtt{up}(\mathtt{s_2})\circ(\neg\mathtt{light}\circ\neg\mathtt{up}(\mathtt{s_1}))$, say, represent identical states. Moreover, double application of function $\neg$ should always neutralize. This intuition is formally captured by stipulating the following formal properties of our special functions:

$$\begin{array}{llcll}
\forall x,y,z. & (x \circ y) \circ z & = & x \circ (y \circ z) & \text{(associativity)} \\
\forall x,y. & x \circ y & = & y \circ x & \text{(commutativity)} \\
\forall x. & x \circ \emptyset & = & x & \text{(unit element)} \\
\forall x. & \neg\neg x & = & x & \text{(double negation)}
\end{array}$$

where the special constant $\emptyset$ denotes a unit element for function $\circ$. This constant will prove useful as formal counterpart of the empty collection of fluent literals. The four axioms (AC1N, for short) constitute an equational theory. Given the law of associativity, from now on we omit parentheses on the level of $\circ$. Notice that the axioms AC1N formalize essential properties of the mathematical concept of a set: The order in which set elements are enumerated is irrelevant, too. From this perspective, constant $\emptyset$ closely corresponds to the empty set.[19] For formal reasons, we introduce a function τ which maps finite sets of fluent expressions $A = \{\ell_1, \ldots, \ell_n\}$ to their term representation $\tau_A = \ell_1 \circ \cdots \circ \ell_n$ (including $\tau_{\{\}} = \emptyset$).

In conjunction with the standard axioms of equality, equational theory AC1N entails the equivalence of two state terms whenever they are built up from an identical collection of fluent literals. By standard equality axioms we mean the following collection:

$$\begin{aligned}
&x = x \\
&x = y \supset y = x \\
&x = y \wedge y = z \supset x = z \\
&x_i = y \supset f(x_1, \ldots, x_i, \ldots, x_n) = f(x_1, \ldots, y, \ldots, x_n) \\
&x_i = y \supset [P(x_1, \ldots, x_i, \ldots, x_n) \equiv P(x_1, \ldots, y, \ldots, x_n)]
\end{aligned} \tag{2.18}$$

for each n-place function symbol f and predicate P $(n \geq 1)$, and for each $1 \leq i \leq n$. All variables are universally quantified.

AC1N plus these standard axioms alone, however, will not suffice for an axiomatization of our action theory. For it will also be necessary to prove unequal two state terms which do not contain the same fluent literals. That is, denials of equalities such as $\texttt{up(s}_1\texttt{)} \circ \texttt{light} \neq \texttt{light} \circ \texttt{up(s}_2\texttt{)}$ need also be derivable. Notice that this inequation is not entailed by the above axioms, for this would require the basic inequality $\texttt{s}_1 \neq \texttt{s}_2$, which is nowhere granted. The standard so-called assumption of unique names provides us with these basic inequalities by means of additional axioms stating that syntactically different terms are never equal. The presence of an equational theory necessitates a weakened version of this assumption. For $\texttt{up(s}_1\texttt{)} \circ \texttt{light}$ and $\texttt{light} \circ \texttt{up(s}_1\texttt{)}$, say, *are* syntactically different but considered equal by our axioms AC1N. A general concept known as *unification completeness* serves this purpose. For a formal introduction to unification complete theories see Annotation 2.1. The set of axioms which constitute an AC1N-unification complete theory allows to derive inequality of state terms not representing the identical collection of fluent literals. These axioms, which include AC1N and the standard axioms of equality, shall be called *extended unique name assumption*, abbreviated *EUNA*.

[19] The reader may wonder why we do not additionally require function $\circ$ to be idempotent, i.e., $\forall x.\ x \circ x = x$, which the comparison with sets would naturally suggest. The (subtle) reason for this decision is given below, right after Proposition 2.9.1.

The formal definition of unification completeness requires some basic notions related to unification wrt. equational theories. Let E be an equational theory, i.e., a set of universally quantified equations. Two terms s and t are said to be *E-equal*, written $s =_E t$, iff $s = t$ is entailed by E plus the standard equality axioms (2.18). A substitution σ is called *E-unifier* of s and t iff $s\sigma =_E t\sigma$. A set $cU_E(s,t)$ of E-unifiers of s and t is called *complete* if it contains, for each E-unifier of s and t, a more or equally general substitution. That is to say, for each substitution σ such that $s\sigma =_E t\sigma$ there exists some $\sigma' \in cU_E(s,t)$ with $(\sigma' \leq_E \sigma)|_{\mathcal{V}ar\,(s) \cup \mathcal{V}ar\,(t)}$. Here, $\mathcal{V}ar\,(t)$ denotes the set of variables occurring in term t, and $(\sigma' \leq_E \sigma)|_V$ means the existence of a substitution θ such that $(\sigma'\theta =_E \sigma)|_V$. The latter holds iff for each variable $x \in V$ the two terms $(x\sigma')\theta$ and $x\sigma$ are E-equal. A *unification complete theory* wrt. E is a consistent set of formulas E^* containing the following:

1. The axioms in E.
2. The standard equality axioms (2.18).
3. Equational formulas, i.e., formulas with " $=$ " as the only predicate, such that for any two terms s and t with variables $\overline{x}$ the following holds:
   a) If s and t are not E-unifiable, then $E^* \models \neg\exists\overline{x}.\ s = t$.
   b) If s and t are E-unifiable, then for each complete set of E-unifiers $cU_E(s,t)$,

$$E^* \models \forall\overline{x} \left[s = t \supset \bigvee_{\sigma \in cU_E(s,t)} \exists\overline{y}.\ \sigma_= \right] \tag{2.19}$$

   where $\overline{y}$ denotes the variables which occur in $\sigma_=$ but not in $\overline{x}$. By $\sigma_=$ we denote the equational formula $x_1 = t_1 \wedge \ldots \wedge x_n = t_n$ constructed from substitution $\sigma = \{x_1 \mapsto t_1, \ldots, x_n \mapsto t_n\}$. If σ is the empty substitution, then $\sigma_=$ evaluates to formula $\top$.

Unification wrt. theory AC1N is well-understood. Whether two terms are AC1N-unifiable is decidable in general, and unifiable terms always admit a finite (but not necessarily singleton) complete set of unifiers. Several complete AC1-unification algorithms are known. Extension to AC1N is straightforward. A unification complete theory AC1N* can be obtained by computing, for any two terms s, t, some complete set $cU_{\text{AC1N}}(s,t)$ of AC1N-unifiers and adding the corresponding equational formula which is to the right of the entailment symbol in (2.19). As a small example, consider the terms $\mathtt{up}(x) \circ z$ and $\mathtt{up(s_1)} \circ \mathtt{up(s_2)} \circ \neg\mathtt{light}$. The set $\{\{x \mapsto \mathtt{s_1}, z \mapsto \mathtt{up(s_2)} \circ \neg\mathtt{light}\}, \{x \mapsto \mathtt{s_2}, z \mapsto \mathtt{up(s_1)} \circ \neg\mathtt{light}\}\}$ is a complete set of AC1N-unifiers of these terms. According to (2.19), axioms AC1N* therefore entail

$$\begin{aligned}\forall x, z\ [\ &\mathtt{up}(x) \circ z = \mathtt{up(s_1)} \circ \mathtt{up(s_2)} \circ \neg\mathtt{light} \supset \\ &x = \mathtt{s_1} \wedge z = \mathtt{up(s_2)} \circ \neg\mathtt{light} \vee x = \mathtt{s_2} \wedge z = \mathtt{up(s_1)} \circ \neg\mathtt{light}\ ]\end{aligned}$$

Annotation 2.1. Unification completeness.

Before entering the axiomatization of our action theory, we prove some crucial properties of *EUNA*. These properties will be exploited later on to model, on the term level, the subset relation and the two set operations difference and union.

Proposition 2.9.1. *Let A and B be two sets of fluent literals.*

1. *If $A \subseteq B$, then $EUNA \models \exists z.\, \tau_A \circ z = \tau_B$, else $EUNA \models \forall z.\, \tau_A \circ z \neq \tau_B$.*
2. *If $A \subseteq B$, then $EUNA \models \forall z\,[\,\tau_A \circ z = \tau_B \;\equiv\; z = \tau_{B \setminus A}\,]$.*
3. *If $A \cap B = \{\}$, then $EUNA \models \forall z\,[\,z = \tau_A \circ \tau_B \;\equiv\; z = \tau_{A \cup B}\,]$.*

Proof.

1. Suppose $A \subseteq B$ and let $Z = B \setminus A$. Then $\tau_A \circ \tau_Z$ and τ_B are AC1N-equal. Since *EUNA* includes AC1N, it entails $\tau_A \circ \tau_Z = \tau_B$, hence also $\exists z.\, \tau_A \circ z = \tau_B$. Suppose $A \not\subseteq B$, then $\tau_A \circ z$ and τ_B are not AC1N-unifiable. According to clause 3a in Annotation 2.1, this implies *EUNA* entails $\forall z.\, \tau_A \circ z \neq \tau_B$.
2. Let z be an arbitrary term, then $\tau_A \circ z$ and τ_B are AC1N-equal iff each fluent literal occurring in $\tau_A \circ z$ also occurs in τ_B and vice versa and no fluent literal occurs twice or more in $\tau_A \circ z$. This in turn is equivalent to z and $\tau_{B \setminus A}$ being AC1N-equal (given that $A \subseteq B$), hence is equivalent to $z = \tau_{B \setminus A}$ under *EUNA*.
3. An arbitrary term z and the term $\tau_A \circ \tau_B$ are AC1N-equal iff each fluent literal occurring in z also occurs in $\tau_A \circ \tau_B$ and vice versa and no fluent literal occurs twice or more in z (given that $A \cap B = \{\}$). This in turn is equivalent to z and $\tau_{A \cup B}$ being AC1N-equal, hence is equivalent to $z = \tau_{A \cup B}$ under *EUNA*.

Qed.

For illustration, consider $A = \{\mathtt{up(s_2)}\}$, $B = \{\neg\mathtt{up(s_1)}, \mathtt{up(s_2)}, \neg\mathtt{light}\}$, and $C = \{\mathtt{up(s_1)}, \mathtt{light}\}$. Then $A \subseteq B$ and $A \cap C = \{\}$. Accordingly, *EUNA* entails each of the following.

$$\exists z.\, \mathtt{up(s_2)} \circ z = \neg\mathtt{up(s_1)} \circ \mathtt{up(s_2)} \circ \neg\mathtt{light}$$
$$\forall z\,[\,\mathtt{up(s_2)} \circ z = \neg\mathtt{up(s_1)} \circ \mathtt{up(s_2)} \circ \neg\mathtt{light} \;\equiv\; z = \neg\mathtt{up(s_1)} \circ \neg\mathtt{light}\,]$$
$$\forall z\,[\,z = \mathtt{up(s_2)} \circ \mathtt{up(s_1)} \circ \mathtt{light} \;\equiv\; z = \mathtt{up(s_1)} \circ \mathtt{up(s_2)} \circ \mathtt{light}\,]$$

Clause 2 of the proposition will be particularly useful when axiomatizing the application of action laws: Suppose C be the condition of some action law, S be some state, and z be a term such that $EUNA \models \tau_C \circ z = \tau_S$, then we know that z represents the set $S \setminus C$. Precisely this is the reason why idempotency of function $\circ$ is not stipulated. For if it would be, then $\tau_C \circ z = \tau_S$ would not imply $z = \tau_{S \setminus C}$. Rather for any set A we would have $\tau_A \circ \tau_A = \tau_A$. But clearly $\tau_A \neq \tau_{A \setminus A}$ unless $A = \{\}$. In contrast, *EUNA* as it stands entails $\tau_A \circ \tau_A \neq \tau_A$ except for $\tau_A = \emptyset$, i.e., $A = \{\}$.

With the extended unique name assumption, *EUNA*, and its properties we are prepared for defining the circumstances under which a term represents a state. In what follows, we tacitly assume a fixed set of entities and fluent names. We use a many-sorted logic language with five sorts, namely, entities, fluent literals, collections of fluent literals, actions, and sequences of actions. Collections are composed of fluent literals, constant $\emptyset$, and our connection function $\circ$. Below, variables of sort entity are indicated by x, variables of sort fluent literal by ℓ, variables of sort action name by a, and variables of sort action sequence by a^*, sometimes with sub- or superscripts. All other variables are of sort collection. Free variables are implicitly assumed universally quantified.

To begin, the following axiom defines a predicate $Holds(\ell, s)$ with the intended meaning that ℓ is contained in s:

$$Holds(\ell, s) \equiv \exists z.\ \ell \circ z = s \tag{2.20}$$

Let, for instance, $s = \neg\mathtt{up(s_1)} \circ \mathtt{up(s_2)} \circ \mathtt{light}$, then $Holds(\mathtt{up(s_2)}, s)$ and $\neg Holds(\mathtt{up(s_1)}, s)$. The next axiom determines the constitutional properties of state terms.

$$State(s) \equiv \forall \ell\,[\,Holds(\ell, s) \equiv \neg Holds(\neg\ell, s)\,] \wedge \forall \ell, z.\ s \neq \ell \circ \ell \circ z \tag{2.21}$$

In words, s represents a state if it contains each fluent literal or its negation but not both. Furthermore, no fluent literal may occur twice (or more) in s. For example, term s above satisfies $State(s)$ provided our action domain is based on exactly the three fluents of which s is composed. If so, then we also have $\neg State(\neg\mathtt{up(s_1)} \circ \mathtt{light})$, $\neg State(\neg\mathtt{up(s_1)} \circ \mathtt{up(s_1)} \circ \mathtt{up(s_2)} \circ \mathtt{light})$, and $\neg State(\neg\mathtt{up(s_1)} \circ \neg\mathtt{up(s_1)} \circ \mathtt{up(s_2)} \circ \mathtt{light})$. The following proposition justifies our definition of *State*:

Proposition 2.9.2. *For a collection* s*,* $EUNA, (2.20), (2.21) \models State(s)$ *iff* $EUNA \models s = \tau_S$ *for some state* S*, else* $EUNA, (2.20), (2.21) \models \neg State(s)$.

Proof. We have $EUNA, (2.20), (2.21) \models State(s)$ iff *EUNA* and the axioms (2.20) and (2.21) entail

$$(\,\exists z.\ \ell \circ z = s \equiv \forall z'.\ \neg\ell \circ z' \neq s\,) \wedge \forall z.\ s \neq \ell \circ \ell \circ z \tag{2.22}$$

for each fluent literal ℓ.

"$\Rightarrow$":

Suppose *EUNA* entails formula (2.22) for each fluent literal ℓ. Define $S = \{\ell : EUNA \models \exists z.\ \ell \circ z = s\}$, then entailment of the first conjunct of (2.22) ensures that S is a state. It also ensures that S consists of precisely the fluent literals which occur in s. Entailment of the second conjunct of (2.22) moreover guarantees that no fluent literal occurs twice or more in s. Altogether this implies $EUNA \models s = \tau_S$.

"$\Leftarrow$":

Suppose $EUNA \models s = \tau_S$ for some state S. Then all fluent literals in τ_S occur exactly once in s. Let ℓ be a fluent literal. S being a state implies that $\{\ell\} \subseteq S$ iff $\{\neg\ell\} \not\subseteq S$. Thus, clause 1 of Proposition 2.9.1 ensures that the first conjunct of (2.22) is entailed. Moreover, since s does not contain any literal twice or more, s and $\ell \circ \ell \circ z$ are not AC1N-unifiable. This implies $EUNA \models \forall z.\ s \neq \ell \circ \ell \circ z$ according to clause 3a in Annotation 2.1. Altogether, it follows that *EUNA* entails formula (2.22) for each literal ℓ.

Qed.

Our next concern is to axiomatize the notion of acceptable states wrt. a set of state constraints. Based on the definition of predicate *Holds*, (2.20), the encoding of fluent formulas is straightforward. In order to state that some formula F is true in the state represented by some term s, each fluent literal ℓ occurring in F is replaced by the expression $\mathit{Holds}(\ell, s)$. E.g., state constraint $\mathtt{light} \equiv \mathtt{up(s_1)} \wedge \mathtt{up(s_2)}$ becomes

$$\mathit{Holds}(\mathtt{light}, s) \equiv \mathit{Holds}(\mathtt{up(s_1)}, s) \wedge \mathit{Holds}(\mathtt{up(s_2)}, s)$$

just like state constraint $\forall x\, [\, \mathtt{fleeing}(x) \supset \mathtt{alive}(x) \,]$ becomes

$$\forall x\ [\ \mathit{Holds}(\mathtt{fleeing}(x), s) \supset \mathit{Holds}(\mathtt{alive}(x), s)\ ]$$

For notational convenience we will simply write $\mathit{Holds}(F, s)$ as abbreviation of the formula thus constructed which states that F is true in s. This encoding of fluent formulas is justified by the following proposition.

Proposition 2.9.3. *Let F be a closed fluent formula and S a state. Then $\mathit{EUNA}, (2.20) \models \mathit{Holds}(F, \tau_S)$ if and only if F is true in S, else $\mathit{EUNA}, (2.20) \models \neg \mathit{Holds}(F, \tau_S)$.*

Proof. A fluent literal ℓ is true in S iff $\{\ell\} \subseteq S$. Following clause 2 of Proposition 2.9.1, the latter is equivalent to $\mathit{EUNA} \models \exists z.\, \ell \circ z = \tau_S$, which in turn is equivalent to $\mathit{EUNA}, (2.20) \models \mathit{Holds}(\ell, \tau_S)$ according to axiom (2.20). The claim follows by straightforward induction on the structure of formula F. *Qed.*

In particular, a state term is $\mathit{Acceptable}_s$ if it satisfies the underlying steady state constraints $\mathcal{C}_s$, and it is *Acceptable* if it satisfies the entire state constraints $\mathcal{C}$, that is,

$$\begin{aligned} \mathit{Acceptable}_s(s) &\equiv \bigwedge_{F \in \mathcal{C}_s} \mathit{Holds}(F, s) \\ \mathit{Acceptable}(s) &\equiv \bigwedge_{F \in \mathcal{C}} \mathit{Holds}(F, s) \end{aligned} \tag{2.23}$$

Suppose, for example,

$$\mathit{Acceptable}(s) \equiv [\, \mathit{Holds}(\mathtt{light}, s) \equiv \mathit{Holds}(\mathtt{up(s_1)}, s) \wedge \mathit{Holds}(\mathtt{up(s_2)}, s) \,]$$

then we can conclude, for instance, $\mathit{Acceptable}(\neg \mathtt{up(s_1)} \circ \mathtt{up(s_2)} \circ \neg \mathtt{light})$ and $\neg \mathit{Acceptable}(\mathtt{up(s_1)} \circ \neg \mathtt{up(s_2)} \circ \mathtt{light})$.

2.9.2 Axiomatizing Action Laws and Causal Relationships

Having defined a suitable representation of states, we now axiomatize two ways of state modification, namely, by means of action laws and causal relationships. To begin, let

$$\{ \alpha_1[\overline{x}_1] = a_1 \ \underline{\text{transforms}}\ C_1 \ \underline{\text{into}}\ E_1, \ldots, \alpha_n[\overline{x}_n] = a_n \ \underline{\text{transforms}}\ C_n \ \underline{\text{into}}\ E_n \}$$

be the underlying set of action laws. These laws define the ternary predicate $Action(a, c, e)$, an instance of which is true if and only if there exists an action law for action a with condition τ_c^{-1} and effect τ_e^{-1}, that is,

$$Action(a, c, e) \ \equiv \ \bigvee_{i=1}^{n} \exists \overline{x}_i \, [\, a = a_i \, \wedge \, c = \tau_{C_i} \, \wedge \, e = \tau_{E_i} \,] \tag{2.24}$$

E.g., the familiar two action laws for toggling a switch may be encoded as

$$\begin{aligned} Action(a, c, e) \ \equiv \ \exists x \, [\, a = \mathtt{toggle}(x) \wedge (\ & c = \neg \mathtt{up}(x) \wedge e = \mathtt{up}(x) \\ \vee\ & c = \mathtt{up}(x) \wedge e = \neg \mathtt{up}(x) \ )\,] \end{aligned}$$

which is a slightly more compact equivalent of the schematic instantiation of (2.24).

Applicability and application of action laws are axiomatized with the help of the properties of *EUNA* elaborated in the previous section. An action law whose condition is represented by collection c is applicable to a state represented by term s if and only if some term z can be found such that $c \circ z$ equals s under *EUNA*. For the latter has been proved equivalent to set τ_c^{-1} being subset of τ_s^{-1}. Entailment of $c \circ z = s$ also guarantees that z contains all fluent literals in s but not in c. A preliminary successor, or rather a representation thereof, is then obtained by concatenating z and the effect term e of the action law in question, yielding term $z \circ e$. Correctness of this axiomatization of applying action laws is guaranteed by the following proposition.

Proposition 2.9.4. *Let $\mathcal{A}$ be a set of action names, $\mathcal{L}$ a set of action laws, S a state, and a an action. Furthermore, let s' be a collection of fluent literals. Then*

$$EUNA, (2.24) \ \models \ \exists c, e, z \, [\, Action(a, c, e) \wedge c \circ z = s \wedge s' = z \circ e \,] \tag{2.25}$$

iff there exists a preliminary successor S' of S and action a such that $EUNA \models s' = \tau_{S'}$, else

$$EUNA, (2.24) \ \models \ \neg \exists c, e, z \, [\, Action(a, c, e) \wedge c \circ z = s \wedge s' = z \circ e \,]$$

Proof. Formula (2.24) for predicate *Action* in conjunction with the standard equality axioms in *EUNA* imply that (2.25) holds iff $\mathcal{L}$ contains an action law instance a <u>transforms</u> C <u>into</u> E such that

$$EUNA \ \models \ \exists z \, [\, \tau_C \circ z = s \wedge s' = z \circ \tau_E \,]$$

This in turn holds iff

1. $C \subseteq S$ (according to clause 1 of Proposition 2.9.1), and
2. $EUNA \models s' = \tau_{(S \setminus C) \cup E}$ (according to clauses 2 and 3 of Proposition 2.9.1 given that $(S \setminus C) \cap E = \{\}$, which is due to $\|C\| = \|E\|$).

These conditions are equivalent to action law a transforms C into E being applicable to S and resulting in $S' = (S \setminus C) \cup E$ where $EUNA \models s' = \tau_{S'}$; hence, the conditions are equivalent to S' being a preliminary successor of S and a. *Qed.*

As an example, consider the state $\{\neg\texttt{up(s}_1\texttt{)}, \texttt{up(s}_2\texttt{)}, \neg\texttt{light}\}$ and the action law instance $\texttt{toggle(s}_1\texttt{)}$ transforms $\{\neg\texttt{up(s}_1\texttt{)}\}$ into $\{\texttt{up(s}_1\texttt{)}\}$. Then *EUNA* entails

$$\exists z\, [\neg\texttt{up(s}_1\texttt{)} \circ z = \neg\texttt{up(s}_1\texttt{)} \circ \texttt{up(s}_2\texttt{)} \circ \neg\texttt{light} \wedge s' = z \circ \texttt{up(s}_1\texttt{)}]$$

with $s' = \texttt{up(s}_1\texttt{)} \circ \texttt{up(s}_2\texttt{)} \circ \neg\texttt{light}$, because z can be substituted by the term $\texttt{up(s}_2\texttt{)} \circ \neg\texttt{light}$. Notice that s' represents the expected preliminary successor state.

Subsequent to the generation of some preliminary successor is the (possibly repeated) application of causal relationships. Let the underlying set of steady relationships be

$$\{\, r_1[\overline{x}_1] = \varepsilon_1 \ \underline{\text{causes}}\ \varrho_1 \ \underline{\text{if}}\ \Phi_1, \ldots, r_m[\overline{x}_m] = \varepsilon_m \ \underline{\text{causes}}\ \varrho_m \ \underline{\text{if}}\ \Phi_m \,\}$$

These relationships define the ternary predicate $Causal_s(\ell_\varepsilon, \ell_\varrho, s)$, an instance of which is true if and only if there exists a steady causal relationship with triggering effect ℓ_ε and ramification ℓ_ϱ and whose context holds in state s, that is,

$$Causal_s(\ell_\varepsilon, \ell_\varrho, s) \;\equiv\; \bigvee_{i=1}^{m} \exists \overline{x}_i\, [\, \ell_\varepsilon = \varepsilon_i \wedge \ell_\varrho = \varrho_i \wedge Holds(\Phi_i, s)\,] \tag{2.26}$$

Likewise, let $\{r_1[\overline{x}_1] = \varepsilon_1$ causes ϱ_1 if $\Phi_1, \ldots, r_n[\overline{x}_n] = \varepsilon_n$ causes ϱ_n if $\Phi_n\}$ be the underlying entire set of causal relationships, which define predicate *Causal* as follows:

$$Causal(\ell_\varepsilon, \ell_\varrho, s) \;\equiv\; \bigvee_{i=1}^{n} \exists \overline{x}_i\, [\, \ell_\varepsilon = \varepsilon_i \wedge \ell_\varrho = \varrho_i \wedge Holds(\Phi_i, s)\,] \tag{2.27}$$

E.g., recall the following familiar four (non-steady) causal relationships.

$$\begin{array}{ll} \texttt{up(s}_1\texttt{)}\ \underline{\text{causes}}\ \texttt{light}\ \underline{\text{if}}\ \texttt{up(s}_2\texttt{)} & \neg\texttt{up(s}_1\texttt{)}\ \underline{\text{causes}}\ \neg\texttt{light}\ \underline{\text{if}}\ \top \\ \texttt{up(s}_2\texttt{)}\ \underline{\text{causes}}\ \texttt{light}\ \underline{\text{if}}\ \texttt{up(s}_1\texttt{)} & \neg\texttt{up(s}_2\texttt{)}\ \underline{\text{causes}}\ \neg\texttt{light}\ \text{if}\ \top \end{array}$$

These are encoded as

$$\begin{aligned} Causal(\ell_\varepsilon, \ell_\varrho, s) \equiv\ & \ell_\varepsilon = \texttt{up(s}_1\texttt{)} \wedge \ell_\varrho = \texttt{light} \wedge Holds(\texttt{up(s}_2\texttt{)}, s) \\ \vee\ & \ell_\varepsilon = \texttt{up(s}_2\texttt{)} \wedge \ell_\varrho = \texttt{light} \wedge Holds(\texttt{up(s}_1\texttt{)}, s) \\ \vee\ & \ell_\varepsilon = \neg\texttt{up(s}_1\texttt{)} \wedge \ell_\varrho = \neg\texttt{light} \\ \vee\ & \ell_\varepsilon = \neg\texttt{up(s}_2\texttt{)} \wedge \ell_\varrho = \neg\texttt{light} \end{aligned} \tag{2.28}$$

Application of (steady) causal relationships to state-effect pairs is axiomatized by defining the predicates $Causes_s(s,e,s',e')$ and $Causes(s,e,s',e')$, respectively, an instance of which is true iff some (steady) causal relationship is applicable to (s,e) and yields (s',e'):

$$Causes_s(s,e,s',e') \equiv \exists \ell_\varepsilon, \ell_\varrho \left\{ \begin{array}{c} Causal_s(\ell_\varepsilon, \ell_\varrho, s) \wedge \exists v.\, \varepsilon_i \circ v = e \\ \wedge \\ \exists z\,[\neg\varrho_i \circ z = s \wedge s' = z \circ \varrho_i] \\ \wedge \\ \left[\begin{array}{c} \forall w.\, \neg\varrho_i \circ w \neq e \wedge e' = e \circ \varrho_i \\ \vee \\ \exists w\,[\neg\varrho_i \circ w = e \wedge e' = w \circ \varrho_i] \end{array} \right] \end{array} \right\} \tag{2.29}$$

This definition needs explanation. Let ε causes ϱ if Φ be some steady causal relationship whose context, Φ, holds in state s. Furthermore, let S, E, S', E' be the sets of fluent literals represented by s, e, s', e'. Then the equational formula in the first row on the right hand side of the equivalence encodes condition $\varepsilon \in E$. The second row simultaneously models the two conditions $\neg\varrho \in S$ and $S' = (S \setminus \{\neg\varrho\}) \cup \{\varrho\}$. Finally, axiomatizing the condition that $E' = (E \setminus \{\neg\varrho\}) \cup \{\varrho\}$ requires case analysis: If $\neg\varrho \notin E$, then we just have to add ϱ to the corresponding term e (third row). If, on the other hand, $\neg\varrho \in E$, then we have to additionally remove the sub-term $\neg\varrho$ from e (fourth row). The definition of *Causes* is analogous but with *Causal* replacing $Causal_s$.

$$Causes(s,e,s',e') \equiv \exists \ell_\varepsilon, \ell_\varrho \left\{ \begin{array}{c} Causal(\ell_\varepsilon, \ell_\varrho, s) \wedge \exists v.\, \varepsilon_i \circ v = e \\ \wedge \\ \exists z\,[\neg\varrho_i \circ z = s \wedge s' = z \circ \varrho_i] \\ \wedge \\ \left[\begin{array}{c} \forall w.\, \neg\varrho_i \circ w \neq e \wedge e' = e \circ \varrho_i \\ \vee \\ \exists w\,[\neg\varrho_i \circ w = e \wedge e' = w \circ \varrho_i] \end{array} \right] \end{array} \right\} \tag{2.30}$$

Correctness of this axiomatization of causal relationships is given by the following proposition.

Proposition 2.9.5. *Let $\mathcal{R}$ be a set of causal relationships. Furthermore, let S be a state, E a set of fluent literals, and s', e' two collections of fluent literals. Then*

$$EUNA, (2.20), (2.27), (2.30) \models Causes(\tau_S, \tau_E, s', e') \tag{2.31}$$

iff there exist two sets of fluent literals S', E' such that $(S,E) \leadsto_{\mathcal{R}} (S',E')$ and $EUNA \models s' = \tau_{S'} \wedge e' = \tau_{E'}$, else

$$EUNA, (2.20), (2.27), (2.30) \models \neg Causes(\tau_S, \tau_E, s', e')$$

Proof. Axioms (2.27) and (2.30) imply that (2.31) holds iff $\mathcal{R}$ contains a causal relationship instance ε $\underline{\text{causes}}$ ϱ $\underline{\text{if}}$ Φ such that the conjunction on the right hand side of the equivalence in (2.30) is entailed. This in turn holds iff

1. $\Phi \wedge \neg\varrho$ is true in S (according to Proposition 2.9.3);
2. $s' = \tau_{(S \setminus \{\neg\varrho\}) \cup \{\varrho\}}$ under *EUNA* (according to clauses 2 and 3 of Proposition 2.9.1);
3. $\{\varepsilon\} \subseteq E$ (according to clause 1 of Proposition 2.9.1); and
4. a) either $\{\neg\varrho\} \not\subseteq E$ and $e' = \tau_{E \cup \{\varrho\}}$ under *EUNA* (according to clauses 1 and 3 of Proposition 2.9.1),
   b) or else $\{\neg\varrho\} \subseteq E$ and $e' = \tau_{(E \setminus \{\neg\varrho\}) \cup \{\varrho\}}$ under *EUNA* (according to clauses 1, 2, and 3 of Proposition 2.9.1).

These conditions are equivalent to causal relationship ε $\underline{\text{causes}}$ ϱ $\underline{\text{if}}$ Φ being applicable to (S, E) and resulting in (S', E') such that $EUNA \models s' = \tau_{S'}$ and $EUNA \models e' = \tau_{E'}$. *Qed.*

As an example, recall (2.28), where *Causal* is defined on the basis of the four causal relationships which describe the usual relation between the switches $\mathtt{s_1}, \mathtt{s_2}$ and the light bulb. Then

$$\begin{aligned} Causes(&\mathtt{up(s_1)} \circ \mathtt{up(s_2)} \circ \neg\mathtt{light}, \mathtt{up(s_1)}, \\ &\mathtt{up(s_1)} \circ \mathtt{up(s_2)} \circ \mathtt{light}, \mathtt{up(s_1)} \circ \mathtt{light}) \end{aligned}$$

as can bee seen by the fact that $\forall w.\ \neg\mathtt{light} \circ w \neq \mathtt{up(s_1)}$ and by replacements $v \mapsto \emptyset$ and $z \mapsto \mathtt{up(s_1)} \circ \mathtt{up(s_2)}$ in (2.30). Notice that this corresponds to $\mathtt{up(s_1)}$ $\underline{\text{causes}}$ $\mathtt{light}$ $\underline{\text{if}}$ $\mathtt{up(s_2)}$ applied to the pair,

$$(\{\mathtt{up(s_1)}, \mathtt{up(s_2)}, \neg\mathtt{light}\}, \{\mathtt{up(s_1)}\})$$

yielding

$$(\{\mathtt{up(s_1)}, \mathtt{up(s_2)}, \mathtt{light}\}, \{\mathtt{up(s_1)}, \mathtt{light}\})$$

Successor states result from repeated application of causal relationships to some preliminary successor. The procedure is to first apply steady relationships until a state obtains that satisfies all steady state constraints. Thereafter, a single arbitrary relationship is applied, followed again by a series of steady relationships, and so on until an overall acceptable state results. The axiomatization of this ramification procedure is based on defining two predicates $Ramify_s(s, e, s', e')$ and $Ramify(s, e, s')$. A valid instance of the former shall indicate the existence of a sequence of steady causal relationships whose application to the state-effect pair (s, e) yields (s', e') such that $Acceptable_s(s')$ holds. Likewise, a valid instance of the latter shall indicate that, for some e', pair (s', e') results from the aforementioned alternated application of steady and arbitrary relationships such that $Acceptable(s')$ holds. In essence the two definitions are transitive closures. As this mathematical concept cannot be expressed in first-order logic, we use the standard way of encoding transitive closure by means of second-order formulas:

$$
\begin{aligned}
&\mathit{Ramify}_s(s, e, s', e') \equiv \\
&\quad \forall \Pi \left\{ \begin{array}{c} \forall s_1, e_1.\ \Pi(s_1, e_1, s_1, e_1) \\ \wedge \\ \left[\begin{array}{l} \forall s_1, e_1, s_2, e_2, s_3, e_3 \\ \quad [\,\Pi(s_1, e_1, s_2, e_2) \wedge \mathit{Causes}_s(s_2, e_2, s_3, e_3) \\ \qquad \supset \Pi(s_1, e_1, s_3, e_3)\,] \end{array} \right] \\ \supset \\ \Pi(s, e, s', e') \end{array} \right\} \\
&\quad \wedge\ \mathit{Acceptable}_s(s')
\end{aligned}
\tag{2.32}
$$

That is, an instance $\mathit{Ramify}_s(s, e, s', e')$ holds if and only if (s, e, s', e') belongs to the transitive closure of Causes_s and if s' satisfies the underlying steady state constraints. Analogously, we define

$$
\begin{aligned}
&\mathit{Ramify}(s, e, s') \equiv \\
&\quad \forall \Pi \left\{ \begin{array}{c} \forall s_1, e_1, s_2, e_2.\ \mathit{Ramify}_s(s_1, e_1, s_2, e_2) \supset \Pi(s_1, e_1, s_2, e_2) \\ \wedge \\ \left[\begin{array}{l} \forall s_1, e_1, s_2, e_2, s_3, e_3, s_4, e_4 \\ \ [\,\Pi(s_1, e_1, s_2, e_2) \wedge \mathit{Causes}(s_2, e_2, s_3, e_3) \\ \ \wedge\, \mathit{Ramify}_s(s_3, e_3, s_4, e_4) \supset \Pi(s_1, e_1, s_4, e_4)\,] \end{array} \right] \\ \supset \\ \exists e'.\ \Pi(s, e, s', e') \end{array} \right\} \\
&\quad \wedge\ \mathit{Acceptable}(s')
\end{aligned}
\tag{2.33}
$$

That is, an instance $\mathit{Ramify}(s, e, s')$ holds if and only if there exists some e' such that (s, e, s', e') belongs to the transitive closure of joining Ramify_s to Causes, and if s' satisfies the underlying entire set of state constraints. Combining the axiomatization of action laws with this definition of ramification leads to the following characterization of successor states:

$$
\begin{aligned}
&\mathit{Successor}(s, a, s') \equiv \\
&\quad \exists c, e, z\, [\, \mathit{Action}(a, c, e) \wedge c \circ z = s \wedge \mathit{Ramify}(z \circ e, e, s')\,]
\end{aligned}
\tag{2.34}
$$

To summarize, let $\mathcal{D}$ be a ramification domain, and let then $\mathcal{FC}_\mathcal{D}$ denote the union of EUNA with the definitions of Holds, axiom (2.20); $\mathit{Acceptable}$ and $\mathit{Acceptable}_s$, axiom (2.23); Action, axiom (2.24); Causes_s and Causes, axioms (2.29) and (2.30); Ramify_s and Ramify, axioms (2.32) and (2.33); and $\mathit{Successor}$, axiom (2.34). The axioms $\mathcal{FC}_\mathcal{D}$ provide a correct axiomatization of transition in ramification domains, as the following theorem shows.

Theorem 2.9.1. *Let $\mathcal{D}$ be a ramification domain and $\mathcal{FC}_\mathcal{D}$ its encoding. Furthermore, let S and S' be two states and a an action. Then*

$$\mathcal{FC}_\mathcal{D} \models \mathit{Successor}(s, a, s')$$

iff there is a successor state S' of S and a such that $\mathit{EUNA} \models s' = \tau_{S'}$, else

$$\mathcal{FC}_\mathcal{D} \models \neg \mathit{Successor}(s, a, s')$$

Proof. The claim follows from the previous propositions regarding correctness of the various axioms in $\mathcal{FC}_\mathcal{D}$ and from the fact that the right hand side formulas of Equations (2.32) and (2.33) encode the appropriate transitive closures. *Qed.*

Example 2.9.1. Let $\mathcal{D}$ be the ramification domain modeling the electric circuit consisting of two switches and a light bulb of Fig. 2.3 as in Example 2.2.2. Then $\mathcal{FC}_\mathcal{D}$ includes the axioms

$$
\begin{aligned}
&Action(a,c,e) \equiv \exists x\,[\, a = \mathtt{toggle}(x) \wedge (\; c = \neg\mathtt{up}(x) \wedge e = \mathtt{up}(x) \\
&\qquad\qquad\qquad\qquad\qquad\qquad \vee\; c = \mathtt{up}(x) \wedge e = \neg\mathtt{up}(x)\;)\,] \\
&Acceptable_s(s) \equiv \top \\
&Acceptable(s) \equiv (Holds(\mathtt{light}, s) \equiv Holds(\mathtt{up}(\mathtt{s_1}), s) \wedge Holds(\mathtt{up}(\mathtt{s_2}), s)) \\
&Causal_s(\ell_\varepsilon, \ell_\varrho, s) \equiv \bot \\
&Causal(\ell_\varepsilon, \ell_\varrho, s) \equiv \quad \ell_\varepsilon = \mathtt{up}(\mathtt{s_1}) \wedge \ell_\varrho = \mathtt{light} \wedge Holds(\mathtt{up}(\mathtt{s_2}), s) \\
&\qquad\qquad\qquad\qquad \vee\; \ell_\varepsilon = \mathtt{up}(\mathtt{s_2}) \wedge \ell_\varrho = \mathtt{light} \wedge Holds(\mathtt{up}(\mathtt{s_1}), s) \\
&\qquad\qquad\qquad\qquad \vee\; \ell_\varepsilon = \neg\mathtt{up}(\mathtt{s_1}) \wedge \ell_\varrho = \neg\mathtt{light} \\
&\qquad\qquad\qquad\qquad \vee\; \ell_\varepsilon = \neg\mathtt{up}(\mathtt{s_2}) \wedge \ell_\varrho = \neg\mathtt{light}
\end{aligned}
$$

We have already seen that $\mathcal{FC}_\mathcal{D}$ entails the following.

$$\exists z\,[\,\neg\mathtt{up}(\mathtt{s_1}) \circ z = \neg\mathtt{up}(\mathtt{s_1}) \circ \mathtt{up}(\mathtt{s_2}) \circ \neg\mathtt{light} \wedge s' = z \circ \mathtt{up}(\mathtt{s_1})\,]$$

It also entails

$$Causes(\mathtt{up}(\mathtt{s_1})\circ\mathtt{up}(\mathtt{s_2})\circ\neg\mathtt{light}, \mathtt{up}(\mathtt{s_1}), \mathtt{up}(\mathtt{s_1})\circ\mathtt{up}(\mathtt{s_2})\circ\mathtt{light}, \mathtt{up}(\mathtt{s_1})\circ\mathtt{light})$$

The latter implies

$$Ramify(\mathtt{up}(\mathtt{s_1}) \circ \mathtt{up}(\mathtt{s_2}) \circ \neg\mathtt{light}, \mathtt{up}(\mathtt{s_1}), \mathtt{up}(\mathtt{s_1}) \circ \mathtt{up}(\mathtt{s_2}) \circ \mathtt{light})$$

according to axioms (2.32) and (2.33) and following the given definitions of *Causes*, $Acceptable_s$, and *Acceptable*. Consequently,

$$Successor(\neg\mathtt{up}(\mathtt{s_1}) \circ \mathtt{up}(\mathtt{s_2}) \circ \neg\mathtt{light}, \mathtt{toggle}(\mathtt{s_1}), \mathtt{up}(\mathtt{s_1}) \circ \mathtt{up}(\mathtt{s_2}) \circ \mathtt{light})$$

according to axiom (2.34). ■

2.9.3 Defining Transition Models

The Fluent Calculus-based axiomatization $\mathcal{FC}_\mathcal{D}$ provides a formal account of the notion of state transition including ramifications. What remains to be done is to formalize the application of whole action sequences to an (unspecified) initial state in view of encoding observations. Our objective is to extend axioms $\mathcal{FC}_\mathcal{D}$ in such a way that there is a one-to-one correspondence between the standard, i.e., 'classical' models of the resulting axiomatization and the models of a set of observations in the sense of our action theory.

The core of this extension consists of two predicates $\mathit{Qualified}(a^*)$ and $\mathit{Result}(a^*, s)$ with the following intuitive meaning. A valid instance of the former indicates that the action sequence a^* is executable starting in the initial state. A valid instance of the latter indicates that s represents the state which would result from performing a^* in the initial state. As a notational convention, if a^* is a (possibly empty) action sequence and a an action, then $[a^*|a]$ shall denote the action sequence which consists in a^* followed by a. The crucial, domain-independent properties of the new predicates are determined by the following axioms:

$$\mathit{Qualified}([\,]) \tag{2.35}$$

$$\mathit{Qualified}([a^*|a]) \equiv \mathit{Qualified}(a^*) \wedge \exists s, s' \begin{bmatrix} \mathit{Result}(a^*, s) \wedge \\ \mathit{Successor}(s, a, s') \end{bmatrix} \tag{2.36}$$

$$\mathit{Result}([a^*|a], s') \supset \forall s\, [\, \mathit{Result}(a^*, s) \supset \mathit{Successor}(s, a, s')\,] \tag{2.37}$$

The reading of the topmost atomic formula, (2.35), is obvious. Formula (2.36) states, in words, that an action sequence $[a^*|a]$ is qualified if so is a^* and if there exists a successor s' of s and a. Finally, implication (2.37) ensures that s' can only be the result of performing the sequence $[a^*|a]$ if s' is a possible successor when executing a in the state resulting from performing a^*. Notice, however, that the axioms (2.36) and (2.37) do not entail the existence of a resulting state whenever the corresponding action sequence is qualified, nor do they entail uniqueness of resulting states. We therefore need to add the following:

$$\exists s.\ \mathit{Result}(a^*, s) \equiv \mathit{Qualified}(a^*) \tag{2.38}$$

$$\mathit{Result}(a^*, s) \wedge \mathit{Result}(a^*, s') \supset s = s' \tag{2.39}$$

Finally, the term intended to represent the initial state should qualify as such, namely, both in being a proper state term and in satisfying all state constraints, that is,

$$\mathit{Result}([\,], s) \supset \mathit{State}(s) \wedge \mathit{Acceptable}(s) \tag{2.40}$$

In what follows, we prove that by introducing these domain-independent axioms we achieve what we have promised. Suppose given a ramification scenario $(\mathcal{O}, \mathcal{D})$ with transition model Σ, and let $\mathcal{FC}_\mathcal{D}$ denote the encoding of domain $\mathcal{D}$ as described in the previous section. Let ι be some classical interpretation of the set of formulas $\mathcal{FC}_\mathcal{D} \cup \{(2.35)\text{–}(2.40)\}$, and let (Σ, Res) be an interpretation for $(\mathcal{O}, \mathcal{D})$. Then we say that ι and (Σ, Res) *correspond* iff for all action sequences a^*, states S, and collections of fluent literals s such that $\mathit{EUNA} \models s = \tau_S$ we find that[20]

[20] Below, by " $[P(t_1, \ldots, t_n)]^\iota$ is true" we mean that the n-tuple $(t_1^\iota, \ldots, t_n^\iota)$ is member of the relation which interpretation ι assigns to predicate P, where t_i^ι $(1 \leq i \leq n)$ denotes the element of ι's universe to which ι maps term t_i.

$$Res(a^*) = S \text{ iff } [Result(a^*, s)]^\iota \text{ is true} \tag{2.41}$$

The notion of correspondence provides the basis for the correctness result of the entire axiomatization.

Theorem 2.9.2. *Let* $(\mathcal{O}, \mathcal{D})$ *be a ramification scenario with transition model* Σ*, and let* $\mathcal{FC}_\mathcal{D}$ *be the axiomatization of* $\mathcal{D}$*. Then for each model* ι *of* $\mathcal{FC}_\mathcal{D} \cup \{(2.35)\text{–}(2.40)\}$ *there exists a corresponding interpretation* (Σ, Res) *for* $(\mathcal{O}, \mathcal{D})$ *and vice versa.*

Proof.
"$\Rightarrow$":
Let ι be a model of $\mathcal{FC}_\mathcal{D} \cup \{(2.35)\text{–}(2.40)\}$. We define a partial mapping Res from finite action sequences to states as follows: $Res(a^*)$ is defined whenever $[Qualified(a^*)]^\iota$ is true; and in case it is defined, let s be a collection of fluent literals such that $[Result(a^*, s)]^\iota$ is true, then $Res(a^*) = S$ for some S such that $EUNA \models s = \tau_S$. By induction on n, we prove that, for any action sequence $a^* = [a_1, \ldots, a_n]$, $Res(a^*)$ satisfies the condition of constituting an interpretation. By construction, the interpretation (Σ, Res) then corresponds to ι.

In the base case, $n = 0$, we have to show that $Res([\,])$ is defined and satisfies the state constraints. According to axiom (2.35), $[Qualified([\,])]^\iota$ is true; hence, $Res([\,])$ is defined. Given that $[Qualified([\,])]^\iota$ is true, axioms (2.38) and (2.39) imply that there is a unique (modulo AC1N) term s such that $[Result([\,], s)]^\iota$ is true. In conjunction with Proposition 2.9.2, formula (2.40) guarantees that s represents a state, which, moreover, satisfies the state constraints according to Proposition 2.9.3.

For the induction step let $n > 0$ and suppose the claim holds for action sequence $[a_1, \ldots, a_{n-1}]$. Let $a^* = [a_1, \ldots, a_{n-1}]$. We have to show that

a) $Res([a^* | a_n])$ is defined iff so is $Res(a^*)$ and $\Sigma(Res(a^*), a_n)$ is not empty, and
b) $Res([a^* | a_n]) \in \Sigma(Res(a^*), a_n)$.

From axiom (2.36) and the induction hypothesis for a^* we conclude that $[Qualified([a^* | a_n])]^\iota$ is true iff $Res(a^*)$ is defined and there exists a term s' such that $[Successor(\tau_{Res(a^*)}, a_n, s')]^\iota$ is true. This proves clause a) according to Theorem 2.9.1. Moreover, axioms (2.38) and (2.39) imply that if $Res([a^* | a_n])$ is defined, then there is a unique (modulo AC1N) term s such that $[Result([a^* | a_n], s)]^\iota$ is true. From axiom (2.37), the induction hypothesis for a^*, and Theorem 2.9.1 it follows that s represents a successor state of $Res(a^*)$ and a_n. This proves clause b).

"$\Leftarrow$":

Let (Σ, Res) be an interpretation for $(\mathcal{O}, \mathcal{D})$, and let ι be an interpretation of $\mathcal{FC}_\mathcal{D} \cup \{(2.35)\text{–}(2.40)\}$ which satisfies the following:

1. ι is a model of $\mathcal{FC}_\mathcal{D}$.

2. For any action sequence a^* and collection of fluent literals s,
   a) $[Qualified(a^*)]^\iota$ is true if and only if $Res(a^*)$ is defined, and
   b) $[Result(a^*, s)]^\iota$ is true if and only if both $Res(a^*)$ is defined and $EUNA \models s = \tau_{Res(a^*)}$.

We have to show that ι is a model of $\mathcal{FC}_\mathcal{D} \cup \{(2.35)\text{–}(2.40)\}$, in which case it corresponds to (Σ, Res) by construction. Given that ι is a model of $\mathcal{FC}_\mathcal{D}$, it suffices to show that it is also a model of axioms (2.35)–(2.40). This in turn can be proved by induction on the length of the argument a^* of *Qualified* and *Result*. This induction proof is completely analogous to the above. *Qed.*

2.9.4 Encoding Observations

The theorem at the end of the previous section shows the adequacy of the domain-independent axioms (2.35)–(2.40) as regards the formal notion of states resulting from performing action sequences. Next, and finally, we concentrate on the scenario-specific expressions that are based on this concept, namely, the observations. Their axiomatization is straightforward. Any observation F after $[a_1, \ldots, a_n]$ is encoded by the formula

$$\exists s\ [\ Result([a_1, \ldots, a_n], s) \wedge Holds(F, s)\ ] \tag{2.42}$$

That is to say, the action sequence $[a_1, \ldots, a_n]$ must admit a resulting state in which, moreover, fluent formula F holds. The addition of these formulas automatically restricts the set of classical models to those which correspond to interpretations in which the respective observations are true. Let $(\mathcal{O}, \mathcal{D})$ be a ramification scenario, then by $\mathcal{FC}_{(\mathcal{O},\mathcal{D})}$ we denote its axiomatization $\mathcal{FC}_\mathcal{D} \cup \{(2.35)\text{–}(2.40)\} \cup \{\bigwedge_{o \in \mathcal{O}} (2.42)\}$. Then classical entailment wrt. this axiomatization and the notion of entailment in our action theory coincide.

Corollary 2.9.1. *Let $(\mathcal{O}, \mathcal{D})$ be a ramification scenario with axiomatization $\mathcal{FC}_{(\mathcal{O},\mathcal{D})}$. An observation F after $[a_1, \ldots, a_n]$ is entailed by $(\mathcal{O}, \mathcal{D})$ iff $\mathcal{FC}_{(\mathcal{O},\mathcal{D})}$ entails the formula $\exists s\, [\, Result([a_1, \ldots, a_n], s) \wedge Holds(F, s)\,]$.*

Proof. Let ι be a model of $\mathcal{FC}_{(\mathcal{O},\mathcal{D})}$ and (Σ, Res) a corresponding interpretation. Given Theorem 2.9.2, it suffices to show that ι is a model of formula (2.42) iff the corresponding observation is true in (Σ, Res). By definition, F after $[a_1, \ldots, a_n]$ is true in (Σ, Res) iff $Res([a_1, \ldots, a_n])$ is defined and F is true in that state. This in turn is equivalent to ι being model of $\exists s\, [\, Result([a_1, \ldots, a_n], s) \wedge Holds(F, s)\,]$ according to axiom (2.38), Proposition 2.9.3, and the fact that ι and (Σ, Res) correspond. *Qed.*

Example 2.9.2. Let $\mathcal{D}$ be the ramification domain modeling the electric circuit consisting of two switches and a light bulb of Fig. 2.3 as in Example 2.2.2. Furthermore, let $\mathcal{FC}_\mathcal{D}$ be its axiomatization as explicated at the end of Section 2.9.2. Let $\mathcal{O}$ consist of the single observation

$$\texttt{light} \ \underline{\text{after}} \ [\texttt{toggle(s}_1\texttt{)}]$$

then $\mathcal{FC}_{(\mathcal{O},\mathcal{D})}$ includes the axiom

$$\exists s \ [\ \mathit{Result}([\texttt{toggle(s}_1\texttt{)}], s) \wedge \mathit{Holds}(\texttt{light}, s)\ ]$$

Let t_s be a term satisfying this conjunction, then axiom (2.37) implies

$$\forall s \ [\ \mathit{Result}([\,], s) \supset \mathit{Successor}(s, \texttt{toggle(s}_1\texttt{)}, t_s)\ ]$$

In turn, this and the fact that $\mathit{Holds}(\texttt{light}, t_s)$ imply

$$\forall s \ [\ \mathit{Result}([\,], s) \supset s = \neg\texttt{up(s}_1\texttt{)} \circ \texttt{up(s}_2\texttt{)} \circ \neg\texttt{light}\ ]$$

according to axiom (2.40) and $\mathcal{FC}_{(\mathcal{O},\mathcal{D})}$. Following axioms (2.35) and (2.38), we thus conclude that

$$\exists s \ [\ \mathit{Result}([\,], s) \wedge \neg\mathit{Holds}(\texttt{up(s}_1\texttt{)}, s) \wedge \mathit{Holds}(\texttt{up(s}_2\texttt{)}, s) \wedge \neg\mathit{Holds}(\texttt{light}, s)\ ]$$

In other words, $(\mathcal{O}, \mathcal{D})$ entails the observation

$$\neg\texttt{up(s}_1\texttt{)} \wedge \texttt{up(s}_2\texttt{)} \wedge \neg\texttt{light} \ \underline{\text{after}} \ [\,]$$

■

2.10 Bibliographic Remarks

The Ramification Problem was introduced and so named in [37].[21] This article also contains the first proposal for a solution, which in essence is a specific realization of globally minimizing change. The authors themselves already argued that this approach may suggest unrealistic indirect effects, but no solution was offered. In [121] the idea was (informally) raised of introducing some notion of preference as regards changes of specific fluents over changes of other fluents. The first formal realization of this approach employed the concept of categorization [65]. The introduction of categories as means to avoid unmotivated indirect effects has later on been used in a variety of frameworks, accompanied by an inventiveness as regards names for these categories—e.g., *frame* vs. *non-frame* fluents [65, 57]; (the latter of which introduces $\mathcal{AR}$ as an extension of the Action Description Language $\mathcal{A}$ mentioned in Section 1.3); *relevant* vs. *dependent* [14]; *persistent* vs. *non-persistent* [18]; *remanent* vs. *dependent* [91]; *inertial* vs. *non-inertial* [41]; or *persistent*, *remanent* and *contingent* fluents [17]. While all being based on the principle of categorization, these approaches differ in the degree of sophistication. The most elementary use of categories is to distinguish two classes of

[21] First published as [36]. The naming was inspired by [28], which was devoted to the problem of how to exploit logical consequences (so-called "ramifications" (*sic*)) of goal specifications in planning problems, with the aim of restricting search space.

fluents, one of which is always subject to the assumption of persistence while the other fluents may vary freely (in view of satisfying the state constraints). More elaborated methods define a partial preference ordering among all possible changes—as we did in Section 2.3. Even further refinement is obtained by state-dependent categorization.[22] The report [92] presented an extension of "Features-and-Fluents" (c.f. Section 1.3) to provide a formal methodology for assessing the range of applicability of different approaches based on categorization.

An early approach which exploits a specific notion of causality to address the Ramification Problem was proposed in [26]. This formalism supports indirect effects deriving from complete descriptions of how the truth-value of a particular fluent might be caused to change. As an example, recall the electric circuit consisting of two switches and a light bulb of Fig. 2.3. The approach of [26] would encode the various possibilities to affect the state of the bulb by these axioms:

$$\begin{aligned} \mathit{Causes}(a,s,\mathtt{light}) \ &\equiv\ \mathit{Causes}(a,s,\mathtt{up(s_1)}) \wedge \mathit{Holds}(\mathtt{up(s_2)},s) \\ &\quad \vee\ \mathit{Causes}(a,s,\mathtt{up(s_2)}) \wedge \mathit{Holds}(\mathtt{up(s_1)},s) \\[6pt] \mathit{Cancels}(a,s,\mathtt{light}) \ &\equiv\ \mathit{Cancels}(a,s,\mathtt{up(s_1)}) \vee \mathit{Cancels}(a,s,\mathtt{up(s_2)}) \end{aligned} \tag{2.43}$$

where $\mathit{Causes}(a,s,f)$ should be read as "executing action a in situation s causes fluent f to become true," $\mathit{Cancels}(a,s,f)$ as "executing action a in situation s causes fluent f to become false," and $\mathit{Holds}(f,s)$ as "fluent f is true in situation s." Suppose we are given a specification of how $\mathtt{up(s_1)}$ may become true (resp. false), namely,

$$\begin{aligned} \mathit{Causes}(a,s,\mathtt{up(s_1)}) \ &\equiv\ a = \mathtt{toggle(s_1)} \wedge \neg \mathit{Holds}(\mathtt{up(s_1)},s) \\ \mathit{Cancels}(a,s,\mathtt{up(s_1)}) \ &\equiv\ a = \mathtt{toggle(s_1)} \wedge \mathit{Holds}(\mathtt{up(s_1)},s) \end{aligned}$$

plus this general axiom meant to express persistence of all non-affected fluents:

$$\begin{aligned} &\mathit{Holds}(f, \mathit{Do}(a,s)) \equiv \\ &\quad \mathit{Causes}(a,s,f) \vee (\, \mathit{Holds}(f,s) \wedge \neg \mathit{Cancels}(a,s,f)\,) \end{aligned} \tag{2.44}$$

where $\mathit{Do}(a,s)$ denotes the situation obtained by executing action a in situation s. One then obtains, e.g., that $\neg \mathit{Holds}(\mathtt{up(s_1)},S_0) \wedge \mathit{Holds}(\mathtt{up(s_2)},S_0) \wedge \neg \mathit{Holds}(\mathtt{light},S_0)$ implies $\mathit{Causes}(\mathtt{toggle(s_1)},S_0,\mathtt{up(s_1)})$ and, hence, also $\mathit{Causes}(\mathtt{toggle(s_1)},S_0,\mathtt{light})$. Thus $\mathit{Holds}(\mathtt{light},\mathit{Do}(\mathtt{toggle(s_1)},S_0))$, as intended. No effort has to be made to suppress an unwanted change of $\mathtt{up(s_2)}$ since no causal relation like the ones in (2.43) exists that may support this. On the other hand, the use of definitional descriptions of causal dependencies, as in (2.43), is restricted to domains where these dependencies are acyclic,

[22] We note, however, that this refinement would not help with our counter-example involving a relay introduced at the end of Section 2.3.

i.e., hierarchical. Otherwise, that is, if fluents depend mutually, unmotivated changes cannot be precluded. To see why, consider the elementary cyclic specification

$$\begin{aligned} Causes(a, s, \mathtt{up}(\mathtt{s_1})) &\equiv Causes(a, s, \mathtt{up}(\mathtt{s_2})) \\ Cancels(a, s, \mathtt{up}(\mathtt{s_1})) &\equiv Cancels(a, s, \mathtt{up}(\mathtt{s_2})) \end{aligned} \tag{2.45}$$

modeling the two switches connected by a tight spring of Fig. 2.5. Suppose both $\mathtt{up}(\mathtt{s_1})$ and $\mathtt{up}(\mathtt{s_2})$ be true in initial situation S_0, and let A be some action whose execution does not affect $\mathtt{up}(\mathtt{s_1})$ nor $\mathtt{up}(\mathtt{s_2})$. Then the two formulas (2.45) in conjunction with the persistence axiom (2.44) are too weak to conclude that $\mathtt{up}(\mathtt{s_1})$ and $\mathtt{up}(\mathtt{s_2})$ remain true. For neither $\neg Cancels(A, S_0, \mathtt{up}(\mathtt{s_1}))$ nor $\neg Cancels(A, S_0, \mathtt{up}(\mathtt{s_2}))$ is entailed.

Cyclic causal dependencies are no obstacle to the approach developed in [14]. This method is based on so-called "causal implications," which, by virtue of being directed, cannot be applied in a non-causal way. For instance, the causal implication $\mathtt{up}(\mathtt{s_1}) \land \mathtt{up}(\mathtt{s_2}) \Rightarrow \mathtt{light}$ has a different meaning than, say, $\mathtt{up}(\mathtt{s_1}) \land \neg\mathtt{light} \Rightarrow \neg\mathtt{up}(\mathtt{s_2})$. In the original approach, successor states are obtained by minimizing change while respecting domain-specific causal implications. A far more simple definition of successor states based on causal implications was presented in [73]. As opposed to [14], this subsequent method moreover allows for the distinction between implicit qualifications and ramifications deriving from state constraints (recall Section 2.8), the necessity of which was first pointed out by [38]. The reason is that in the approach [14] one always strives for a successor state no matter how many changes are necessary to this end, while [73] additionally requires all changes be explicitly grounded on some causal implication. A more detailed comparison between the two approaches can be found in [73]. A closely related approach, [30, 31], is based on a nonmonotonic theory of "conditional entailment" and uses expressions similar to causal implications.

The nature of causal implications resembles the concept of causal relationships propagated in this book. In fact, our fixpoint characterization of causal minimizing-change successors was borrowed from [73] and transferred to causal relationships. A crucial difference between causal implications and relationships is that only the latter distinguishes between a context an explicitly occurred effect. E.g., the two relationships ℓ_1 <u>causes</u> ℓ <u>if</u> ℓ_2 and ℓ_2 <u>causes</u> ℓ <u>if</u> ℓ_1 are not interchangeable while both corresponding to the identical causal implication, viz. $\ell_1 \land \ell_2 \Rightarrow \ell$. The following simple scenario illustrates the usefulness of causal relationships being more expressive in this sense.

Example 2.10.1. Consider a more subtle, ancient method to hunt turkeys, namely, by using a (manually activated) trapdoor. The state of this trapdoor is formalized by the fluent name $\mathtt{trap\text{-}open}^0$. The fluent $\mathtt{at\text{-}trap}(\mathtt{turkey})$ describes whether the victim is in the dangerous zone or not, and the fluent name $\mathtt{alive}^1$ is used as before. The ground underneath the trapdoor is designed such that nothing being $\texttt{at-trap}$ with the trapdoor open can be alive. This is represented by the state constraint

$$\forall x\,[\,\texttt{at-trap}(x) \land \texttt{trap-open} \supset \neg\texttt{alive}(x)\,] \tag{2.46}$$

We can open the trapdoor and entice the turkey, respectively, via these two action laws:

$$\texttt{open}\ \ \underline{\text{transforms}}\ \ \{\neg\texttt{trap-open}\}\ \ \underline{\text{into}}\ \ \{\texttt{trap-open}\}$$
$$\texttt{entice}(x)\ \ \underline{\text{transforms}}\ \ \{\neg\texttt{at-trap}(x)\}\ \ \underline{\text{into}}\ \ \{\texttt{at-trap}(x)\}$$

While the state of the trapdoor can possibly affect the turkey being alive, the animal is alert to the extent that it would never kill itself by moving towards the open trapdoor; hence, $\texttt{trap-open}$ may influence $\texttt{alive(turkey)}$ but $\texttt{at-trap(turkey)}$ may not so. The latter is intended to give rise to the implicit qualification $\neg\texttt{trap-open}$ for the action $\texttt{entice(turkey)}$. Therefore the correct influence information is $\mathcal{I} = \{(\texttt{trap-open}, \texttt{alive(turkey)})\}$. In conjunction with our state constraint (2.46), this determines a single causal relationship, viz.

$$\texttt{trap-open}\ \ \underline{\text{causes}}\ \ \neg\texttt{alive(turkey)}\ \ \underline{\text{if}}\ \ \texttt{at-trap(turkey)} \tag{2.47}$$

Given the state $S = \{\texttt{alive(turkey)}, \texttt{at-trap(turkey)}, \neg\texttt{open}\}$ (say, after having enticed the turkey), performing the action $\texttt{open}$ yields the state $\{\texttt{alive(turkey)}, \texttt{at-trap(turkey)}, \texttt{open}\}$ as preliminary successor, which violates the state constraint. Since $\texttt{trap-open}$ occurred as effect, the given causal relationship is applicable, which results in the unique and expected successor state $\{\neg\texttt{alive(turkey)}, \texttt{at-trap(turkey)}, \texttt{trap-open}\}$.

Consider state $S = \{\texttt{alive(turkey)}, \neg\texttt{at-trap(turkey)}, \texttt{trap-open}\}$ and action $\texttt{entice(turkey)}$, on the other hand. Performing this action yields the preliminary successor $\{\texttt{alive(turkey)}, \texttt{at-trap(turkey)}, \texttt{trap-open}\}$, too, but now obtained through the direct effect $E = \{\texttt{at-trap(turkey)}\}$. Therefore the only available causal relationship is not applicable; hence, no successor state exists. In other words, $\neg\texttt{trap-open}$ is revealed as additional, implicit qualification for $\texttt{entice(turkey)}$, as expected. Notice that we are only able to distinguish between these two cases by employing causal relationship (2.47) but not the analogue, namely,

$$\texttt{at-trap(turkey)}\ \ \underline{\text{causes}}\ \ \neg\texttt{alive(turkey)}\ \ \underline{\text{if}}\ \ \texttt{trap-open}$$

Since these two relationships both correspond to the same causal implication $\texttt{at-trap(turkey)} \land \texttt{trap-open} \Rightarrow \neg\texttt{alive(turkey)}$, this distinction goes beyond the expressiveness of causal implications. ■

In [69], which improves previous work where indirect effects are 'compiled' into action descriptions [68], first-order formulas resembling causal relationships are employed to define dependencies between effects and their indirect consequences. These formulas are of the form

$$\Phi(s) \wedge \mathit{Caused}(f_1, v_1, s) \wedge \ldots \wedge \mathit{Caused}(f_n, v_n, s) \supset \mathit{Caused}(f, v, s) \quad (2.48)$$

where $\mathit{Caused}(f, v, s)$ should be read as "fluent f is caused to take on truth-value v in situation s," and where $\Phi(s)$ describes properties of situation s. Notice the distinction between a context, i.e., Φ, and explicitly occurring effects $f_1, \ldots, f_n$. E.g., a formalization of the electric circuit involving two switches and a light bulb of Fig. 2.3 would include the specification

$$\mathit{Holds}(\mathtt{up(s_2)}, s) \wedge \mathit{Caused}(\mathtt{up(s_1)}, \mathit{True}, s) \supset \mathit{Caused}(\mathtt{light}, \mathit{True}, s) \quad (2.49)$$

along with the action definition[23]

$$\neg \mathit{Holds}(\mathtt{up(s_1)}, s) \supset \mathit{Caused}(\mathtt{up(s_1)}, \mathit{True}, \mathit{Do}(\mathtt{toggle(s_1)}, s)) \quad (2.50)$$

The general axiom of persistence used in this context is

$$\neg \exists v.\, \mathit{Caused}(f, v, \mathit{Do}(a, s)) \supset (\, \mathit{Holds}(f, \mathit{Do}(a, s)) \equiv \mathit{Holds}(f, s)\,) \quad (2.51)$$

This axiom is of course useless unless an instance $\mathit{Caused}(F, V, \mathit{Do}(A, S))$ is provably false whenever it does not follow from the domain-specific axioms both for direct effects of actions (like in (2.50)) and for causal dependencies (like in (2.49)). This is formally achieved by minimizing the set of true instances of *Caused* via so-called circumscription [77]. For instance, suppose a situation S_0 be specified by $\neg \mathit{Holds}(\mathtt{up(s_1)}, S_0) \wedge \mathit{Holds}(\mathtt{up(s_2)}, S_0) \wedge \neg \mathit{Holds}(\mathtt{light}, S_0)$, and let $S_1 = \mathit{Do}(\mathtt{toggle(s_1)}, S_0)$. Then the axioms (2.49) and (2.50) entail $\mathit{Caused}(\mathtt{up(s_1)}, \mathit{True}, S_1)$, hence $\mathit{Caused}(\mathtt{light}, \mathit{True}, S_1)$. They do not entail $\exists v.\, \mathit{Caused}(\mathtt{up(s_2)}, v, S_1)$. Minimizing *Caused* therefore allows to conclude $\neg \exists v.\, \mathit{Caused}(\mathtt{up(s_2)}, v, S_1)$ and, hence, $\mathit{Holds}(\mathtt{up(s_2)}, S_1)$ according to axiom (2.51).

An alternative causality-based method to address the Ramification Problem, which also includes the distinction between context and triggering effect, is [47].

The causality-oriented approaches cited so far all intrinsically follow the policy of minimizing change. This amounts to rejecting any potential successor state whose distance to the original state is strictly greater than the distance of another proper successor state. With our electric circuit involving the light detecting device (Example 2.6.3) we have seen that the applicability of this paradigm is limited. Causal relationships and their successive application to preliminary successor states constitute the first approach that does not follow the principle of minimizing change [113];[24] the notion of steady state constraints has been introduced in [116]. In [93], an extension to "Features-and-Fluents" was developed as an alternative to the aforementioned [92] for categorization-based approaches. The newly proposed action

[23] For the sake of simplicity we neglect the concept of action preconditions here.

[24] First published as [109]. The approach [71], which also supports non-minimal successor states, emerged at the same time but proved erroneous in producing lots of 'magic' indirect effects.

theory is causality-oriented and follows the idea of successively generating indirect effects, starting with preliminary successors and continuing until an acceptable state is obtained. Minimality of changes is not required. This general approach was used in [48] to extend the so-called Temporal Action Logic [23]. The approach [10] builds on the action calculus called Linear Connection Method (see below). It introduces resource-sensitive implications which are syntactically identical to the action descriptions used therein. These additional implications are applied with higher priority and have certain fluent literals marked which must have previously occurred as effects in order for the implication to apply. The resulting calculus can be viewed as realization of causal relationships as means to address the Ramification Problem.

In [21], a least fixpoint semantics is used to model chains of indirect effects. The article also contains the proposal to extend the Ramification Problem to indirect effects which derive from constraints that connect the states in different situations.

An approach which is considerably different from all methods discussed so far yet still related is based on networks representing probabilistic causal theories [81]. These networks describe, in the first place, static dependencies among their components. As argued in [82, 83], however, the truth-values of one or more nodes may be reset dynamically and, then, the values of all depending nodes need to be adjusted according to standard (Bayesian) rules of probability. This can be regarded as generating indirect effects. If probability values are restricted to the binary 0/1-case, then a network whose nodes are fluent names resembles our concept of influence information. For instance, Fig. 2.13a) depicts a network suitable for our electric circuit with the relay and light detector. In view of a general solution to the Ramification Problem, some restrictions of causal networks are worth mentioning. First, we note that the resulting value of a node, after having fixed the direct effects, must not be computed until all new values of its predecessors have been determined. Consequently, the proposition `detect` in the network of Fig. 2.13a) necessarily remains false after toggling switch $\mathtt{s_1}$ in the state depicted in Fig. 2.8; hence, the non-minimal successor state where a light flash has been detected cannot be obtained. Second, recall the domain with the trapdoor, Example 2.10.1. Since fluent `trap-open` changing possibly affects `alive(turkey)`, depending on `at-trap(turkey)`, the adequate network is the one depicted in Fig. 2.13b). This, however, does not allow to distinguish between the two situations where either `trap-open` becomes true with `at-trap(turkey)` being true, or it happens to be the other way round. Hence, the distinction between context and triggering effect is not supported by causal networks. Finally, networks representing causal theories are based on acyclic graphs, which means that cyclic dependencies, like the one given by our switches of Fig. 2.5, which are being connected by a spring, cannot be represented (c.f. Fig. 2.13c)).

The Fluent Calculus paradigm we used for an axiomatization of our action theory emerged from the urge to solve the fundamental so-called Frame

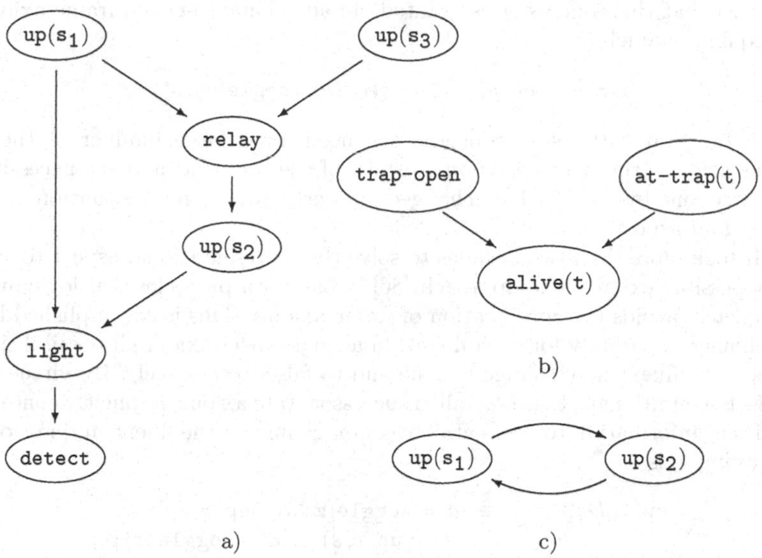

Figure 2.13. Graphs representing the structural dependencies between fluents regarding a) Example 2.6.3, b) Example 2.10.1 (where `t` abbreviates `turkey`), and c) state constraint $\mathtt{up(s_1)} \equiv \mathtt{up(s_2)}$ (inducing a cycle), respectively.

Problem. Introduced in the context of Situation Calculus [74], it concerns the at first glance trivial but in fact highly problematic challenge to specify, in logic, that non-affected fluents keep their truth-values during the performance of actions. To see where the difficulties lie, observe first that naïvely describing the effects of actions such as $\mathtt{toggle}(x)$ by means of classical material implication does not work. E.g., the specification

$$Do(\mathtt{toggle}(x)) \supset (\neg\mathtt{up}(x) \supset \mathtt{up}(x)) \tag{2.52}$$

is inconsistent with, say, $\neg\mathtt{up(s_1)} \wedge Do(\mathtt{toggle(s_1)})$, i.e., the seemingly natural logical formalization of a situation where switch $\mathtt{s_1}$ is down and is about to being toggled. This is why Situation Calculus introduces an additional so-called situation argument to each action and fluent, restricting its scope to a particular one out of many possible situations. The effect description (2.52) thus becomes

$$\neg\mathtt{up}(x, s) \supset \mathtt{up}(x, Do(\mathtt{toggle}(x), s))$$

where s denotes some abstract situation and $Do(\mathtt{toggle}(x), s)$ the successor situation resulting from performing $\mathtt{toggle}(x)$ in s. This representation technique, however, raises the problem of how to conclude that some other fluent which holds in s, say $\mathtt{up(s_2)}$, still holds in situation $Do(\mathtt{toggle(s_1)}, s)$.

In order that this conclusion is granted, an additional so-called frame axiom is required, namely,

$$x \neq y \wedge \mathtt{up}(y, s) \supset \mathtt{up}(y, Do(\mathtt{toggle}(x), s))$$

Now, the Frame Problem concerns the need for a large number of these frame axioms (the *representational* aspect of the problem) and the necessity to carry, one-by-one, each unchanged property to the next situation (the *inferential* aspect).

It took more than two decades to solve the representational aspect to the best possible extent. The approach [86] is based on pure classical logic and completely avoids the specification of frame axioms. This is accomplished by combining, separately for each fluent, in a single effect axiom all possibilities of how the fluent may change to true and to false, respectively. By virtue of being bi-conditionals, these so-called successor state axioms implicitly contain sufficient information to also entail any *non*-change of the fluent in question. An example is

$$\begin{aligned} \mathtt{up}(x, Do(a, s)) \;\equiv\; & a = \mathtt{toggle}(x) \wedge \neg\mathtt{up}(x, s) \\ & \vee \neg\,[\,\mathtt{up}(x, s) \supset a = \mathtt{toggle}(x)\,] \end{aligned}$$

Given $\neg\mathtt{up}(\mathtt{s_1}, S_0) \wedge \mathtt{up}(\mathtt{s_2}, S_0)$, this axiom entails $\mathtt{up}(\mathtt{s_1}, Do(\mathtt{toggle}(\mathtt{s_1}), S_0))$ and also $\mathtt{up}(\mathtt{s_2}, Do(\mathtt{toggle}(\mathtt{s_1}), S_0))$, tacitly assuming that $\mathtt{toggle}(\mathtt{s_1}) \neq \mathtt{toggle}(\mathtt{s_2})$. While the concept of successor state axioms perfectly solves the representational aspect of the Frame Problem, it does not at all address the inferential aspect. For it still requires, for each non-affected fluent, separate application of one of these axioms in order to conclude that the fluent keeps its truth-value in the resulting situation.

The STRIPS framework [27] was an early development in response to the inferential challenge raised by the Frame Problem. STRIPS encodes states as sets of fluents, and the performance of actions is specified operationally, namely, by removal and addition of certain fluents to these sets. Apparently, this avoids investigation of any non-affected fluent. In compensation, the operational, non-declarative nature of this approach causes the loss of both expressiveness and flexibility of logic. With the aim of regaining the latter without losing the computational merits of STRIPS, the *Linear Connection Method* has been developed in [9], a precursor of *Linear Logic* [39]. It employs a non-classical, resource-sensitive implication that allows to specify effects of actions as straightforward as in

$$\neg\mathtt{up}(x) \,\&\, Sit(s) \;\multimap\; \mathtt{up}(x) \,\&\, Sit(Do(\mathtt{toggle}(x), s)) \tag{2.53}$$

where "&" denotes a non-idempotent conjunction, "$\multimap$" the aforementioned, non-classical implication, and $Sit(s)$ should be read as "the current situation is s." Being resource-sensitive, the antecedent of an implication like (2.53) is 'consumed' whenever deriving the consequent. Consequently, $\neg\mathtt{up}(\mathtt{s_1}) \,\&\, \mathtt{up}(\mathtt{s_2}) \,\&\, Sit(S_0)$, say, in conjunction with formula (2.53) entails $\texttt{up}(\texttt{s}_1)\ \&\ \texttt{up}(\texttt{s}_2)\ \&\ \mathit{Sit}(\mathit{Do}(\texttt{toggle}(\texttt{s}_1), S_0))$ but not the contradiction $\neg\texttt{up}(\texttt{s}_1)\ \&\ \texttt{up}(\texttt{s}_2)\ \&\ \mathit{Sit}(S_0)\ \&\ \texttt{up}(\texttt{s}_1)\ \&\ \mathit{Sit}(\mathit{Do}(\texttt{toggle}(\texttt{s}_1), S_0))$. A comparison between Linear Connection Method and Situation Calculus can be found in [8]. A thorough study of the relationship with Linear Logic is reported in [29], where also an inference engine for the Linear Connection Method is described.

The Fluent Calculus paradigm [51], so named by [12], embeds in pure classical logic the notion of resource-sensitivity as means to reflect the dynamics of state transitions. For a specific class of planning problems, axiomatizations based on Linear Connection Method, Fluent Calculus, and Linear Logic (in the variant of [72]) have been proved equivalent [45]. In [118] it is shown that under the provision that actions do not have an unbounded number of direct effects, any Situation Calculus specification along the line of [86] can be transformed into an essentially equivalent Fluent Calculus specification, in which at the same time the representational and the inferential aspect of the Frame Problem are addressed. Axiomatizations using the paradigm of Fluent Calculus have been developed for a variety of ontological aspects—we just mention concurrency of actions [12], continuous change [50, 117], hierarchical planning [25], and nondeterministic [119] and complex actions [52]. Fluent Calculus grounds on the algebraic theory of Abelian monoid, which is given by the axioms of associativity, commutativity, and existence of a unit element. Unification algorithms specialized in this equational theory are described in, e.g., [105, 16, 44] (for an overview see [80]). An analysis of complexity is performed in [56]. The related notion of unification completeness has been developed in the context of logic programming. Introductions to this principle can be found in, e.g., [54, 101, 111]. The existence of a unification complete theory for AC1 has been shown in [53].

3. The Qualification Problem

3.1 Abnormal Action Disqualifications

Suppose you want to start your car to take a ride. What are the preconditions for doing this? Presumably you first want to make sure that you got hold of the ignition key. Does this justify the conclusion that the action can now be successfully performed? Not if you are overly cautious, in which case you might also want to check that the gas tank is not empty. Still, however, there are numerous other imaginable causes for being unable to ignite the engine of your car. For instance, you have failed to make sure that no potato clogs the tail pipe, despite the fact that this necessarily renders the intended action impossible.

Of course there are good reasons for ignoring the possibility of the tail pipe housing a potato. It seems highly inappropriate, in general, to explicitly verify every imaginable precondition of an action—in fact this is even impossible: Apart from the fact that besides a clear tail pipe there are lots of further disqualifying obstacles to our example action, how would you ensure that after checking the tail pipe it does not become clogged during you walking to the front door and taking a seat, prior to trying to ignite the engine? Actors in real world environments would therefore be totally paralyzed would they never perform an action without worrying about every however unlikely obstacle.

The vast majority of preconditions for actions in daily life are so likely to being satisfied that common sense assumes them away as long as there is no evidence to the contrary. On the other hand, ignoring all these 'abnormal' action disqualifications *prima facie* also means to being able to handle situations where the prior assumption of executability turns out wrong. This is in contrast to the idealistic view our action theory takes so far. The existence of a successor state according to the underlying transition model is supposed to guarantee the successful performance of the action in question. Making the contrary observation is considered impossible insofar as it would render inconsistent the entire scenario. Let, for example, the state of a car's engine be denoted by the fluent `runs` and suppose the action of starting the car be specified by the action law `ignite` transforms {¬`runs`} into {`runs`}. Then the two observations ¬`runs` after [ ] and ¬`runs` after [`ignite`] constitute a scenario that admits no models at all.

M. Thielscher: Challenges for Action Theories, LNAI 1775, pp. 85-118, 2000.

Despite being a necessary precondition, the absence of obstacles like "there is a potato in the tail pipe" should, however, not be added to the condition part of the above action law. For otherwise this would require what has been said inappropriate, namely, the inspection of all abnormal disqualifications of this action prior to assuming it executable. The general challenge, therefore, is to weaken the strict conclusion that actions are guaranteed to producing the expected effect once all specified preconditions are satisfied. This conclusion should become an assumption *by default*. As such it is to be made as long as there is no information to the contrary. Developing a formal account of this concept within the framework of an action theory is the *Qualification Problem*. Solving it is necessary in view of applying action theories to real-world environments, which do not conform with the idealistic view in that most if not all actions are potentially subject to abnormal disqualification.

Let us state the problem more precisely. Suppose given the two fluents **runs** and **in(pt)** stating, respectively, whether the engine of the car is running and whether its tail pipe houses a potato. As above, the action **ignite** shall be specified by the action law **ignite** transforms {¬**runs**} into {**runs**}. Consider the situation where the only available information concerns the state of the engine, which is known to be off. Nothing is known as to whether or not the tail pipe is clogged. The formal specification of this scenario thus consists of the observation

¬**runs** after [ ]

Then by default it should follow that **runs** after [**ignite**]. The argument supporting this conclusion would be that there is no hint at a potato being placed in the tail pipe and the only 'regular' precondition of the action, viz. that the engine is not running already, is true in the initial state.[1] Our action is therefore to be assumed executable and so to have the expected effect. But now consider the situation where we do not only know the engine is off but also that there *is* a potato in the tail pipe, that is,

¬**runs** after [ ]
in(pt) after [ ]

Then it should no longer follow that igniting the engine is possible nor, hence, that the observation **runs** after [**ignite**] holds.

This example reveals a crucial principle with which the Qualification Problem is inherently connected. Namely, additional observations, in our case **in(pt)** after [ ], may force the withdrawal of previously valid conclusions. A so-called *nonmonotonic* entailment relation for observations is required to this end. This is in contrast to our current action theory being monotonic in that sense: Whenever two sets of observations $\mathcal{O}_1$ and $\mathcal{O}_2$ satisfy $\mathcal{O}_1 \subseteq \mathcal{O}_2$,

[1] By "regular" preconditions we generally mean both the conditions occurring in an action law and the implicit qualifications deriving from state constraints.

then any observation entailed by $\mathcal{O}_1$ is also entailed by $\mathcal{O}_2$. This is a consequence of the fact that adding observations only reduces the set of models. A fundamental task when addressing the Qualification Problem is therefore to modify an underlying monotonic entailment relation so as to become a nonmonotonic one.

3.2 Minimizing Abnormal Disqualifications

Recall from above the scenario where the only observation concerns the engine, stating that it is not running. No information is given as to whether the tail pipe is clogged or not. According to our action theory, the set of models of this scenario divides into two classes, one of which consists in all models which consider initially false fluent `in(pt)`, while the models of the other class consider that fluent true. Suppose somewhere in the underlying domain specification it is said that `in(pt)` being true renders action `ignite` unexecutable even if all regular preconditions are satisfied. Then half the models entail this very conclusion, i.e., that the action cannot be performed, the others do not so. Hence, if all models need to be taken into account when deciding on entailment, then nothing follows from the given observations as to the action in question being executable. As argued in the introduction to the Qualification Problem, however, it is reasonable to conclude, by default, that starting the engine *is* possible. Granting this conclusion therefore requires to ignore all those models which claim the contrary.

Dubious and unsound as it may seem to simply disregard 'inconvenient' models preventing the desired conclusion there is good reason for so doing in this particular situation. It has been said that an abnormal disqualification should be assumed away by default, that is, as long as this assumption is consistent with the available information. From the model perspective, this amounts to disregarding all models which entail an abnormal action disqualification if only at least one model exists of the scenario at hand where this abnormality does not hold. One way of achieving this formally is to introduce a so-called preference criterion. Being a partial order on models, this criterion will allow us to select the most preferred ones and, then, to confine ourselves to those when talking about entailment.

The concept of model preference is inherently nonmonotonic, just as the Qualification Problem requires it. Additional observations may invalidate some of the models, among which, moreover, might be all previously preferred ones. As a consequence, models that have been disregarded earlier may now come to light and give rise to a very different set of entailed formulas. Recall our example where we add to `¬runs` <u>after</u> [ ] the observation that there is a potato in the tail pipe initially, i.e., `in(pt)` <u>after</u> [ ]. This new information falsifies every model that claims the possibility to ignite the engine. The remaining, previously unpreferred models all suggest an abnormal disqualification of the action, hence do now entail `ignite` <u>inexecutable</u> <u>after</u> [ ], as we will note this kind of observation:

Definition 3.2.1. *Let $\mathcal{E}$, $\mathcal{F}$, and $\mathcal{A}$ be sets of entities, fluent names, and action names, respectively. A* disqualification observation *is of the form*

$$a \ \underline{\text{inexecutable}}\ \underline{\text{after}}\ [a_1, \ldots, a_n]$$

where each of $a, a_1, \ldots, a_n$ is an action ($n \geq 0$). If Res is a partial mapping from finite (possibly empty) action sequences to states, then this disqualification observation is true *in Res iff $Res([a_1, \ldots, a_n])$ is defined but $Res([a_1, \ldots, a_n, a])$ is not.* ■

In what follows, the term "observation" refers both to observations in the original sense as introduced in Chapter 1 (c.f. Definition 1.2.5) and to disqualification observations.

For a formal account of the approach to the Qualification Problem sketched above we need means to connect abnormal disqualifications of actions with the situations that give rise to them. The following notion serves this purpose.

Definition 3.2.2. *Let $\mathcal{E}$, $\mathcal{F}$, and $\mathcal{A}$ be sets of entities, fluent names, and action names, respectively. A* disqualifying condition *is an expression of the form $F \supset disq(a)$ where F is a fluent formula and a an action.* ■

For notational convenience, both F and a may contain variables, in which case the disqualifying condition $F \supset disq(a)$ is regarded as representative for all its ground instances. An example is $\mathtt{in}(x) \supset disq(\mathtt{ignite})$, stating that any object clogging the tail pipe unqualifies the action of starting the engine.

A disqualifying condition $F \supset disq(a)$ indicates that whenever formula F is true, then action a cannot be performed even if all of its regular preconditions are satisfied. Having disqualifying conditions requires an extended notion of interpretations and models (Σ, Res) of action scenarios. If an action a is disqualified in a particular state $Res([a_1, \ldots, a_n])$, then this dictates that $Res([a_1, \ldots, a_n, a])$ is undefined regardless of whether the underlying transition model, Σ, suggests a successor state of $Res([a_1, \ldots, a_n])$ and a. To allow for comparison of models in view of preferring those with the fewest possible abnormal disqualifications, a third component is introduced into both interpretations and models. This new argument, denoted Ab, reflects all situations where an action cannot be performed on account of some disqualifying condition.

Definition 3.2.3. *Let $(\mathcal{O}, \mathcal{D})$ be a ramification scenario, and let $\mathcal{Q}$ be a set of disqualifying conditions. An* interpretation with abnormalities *for $(\mathcal{O}, \mathcal{D})$ is a triple (Σ, Res, Ab) where Σ is the transition model of $\mathcal{D}$, Res is a partial function which maps finite (possibly empty) action sequences to acceptable states, and Ab is a set of non-empty action sequences such that the following holds:*

1. *$Res([\,])$ is defined.*
2. *For any sequence $a^* = [a_1, \ldots, a_{k-1}, a_k]$ of actions ($k > 0$):*

a) $Res(a^*)$ *is defined if and only if* $Res([a_1,\ldots,a_{k-1}])$ *is defined;* $\Sigma(Res([a_1,\ldots,a_{k-1}]),a_k)$ *is not empty; and no* $F \supset disq(a_k) \in \mathcal{Q}$ *exists such that* F *is true in* $Res([a_1,\ldots,a_{k-1}])$.

b) $Res(a^*) \in \Sigma(Res([a_1,\ldots,a_{k-1}]),a_k)$.

c) $a^* \in Ab$ *if and only if* $Res([a_1,\ldots,a_{k-1}])$ *is defined and there is some* $F \supset disq(a_k)$ *such that* F *is true in* $Res([a_1,\ldots,a_{k-1}])$. ■

Put in words, whenever $Res([a_1,\ldots,a_{k-1}])$ entails the antecedent of a disqualifying condition for action a_k, then $Res([a_1,\ldots,a_k])$ is not defined and Ab includes the sequence $[a_1,\ldots,a_k]$.

The appropriate preference criterion and entailment relation can then be defined straightforwardly on the basis of comparing the additional components, Ab. Informally speaking, the less action sequences are declared disqualified by a model the less abnormal the latter. Entailment is then decided on the basis of least abnormal models, which thus are the preferred ones.

Definition 3.2.4. *Let* $(\mathcal{O},\mathcal{D})$ *be a ramification scenario with transition model* Σ *and* $\mathcal{Q}$ *a set of disqualifying conditions. If* $I = (\Sigma, Res, Ab)$ *and* $I' = (\Sigma, Res', Ab')$ *are interpretations with abnormalities for* $(\mathcal{O},\mathcal{D})$*, then* I *is* less abnormal *than* I'*, written* $I \prec I'$*, iff* $Ab \subsetneq Ab'$. *A* model with abnormalities *of* $(\mathcal{O},\mathcal{D})$ *is an interpretation* (Σ, Res, Ab) *such that each* $o \in \mathcal{O}$ *is true in* Res. *A model is* preferred *iff there is no other model which is less abnormal. An observation* o *is* entailed *iff it is true in all preferred models with abnormalities of* $(\mathcal{O},\mathcal{D})$. ■

Let us see how this account of abnormal action disqualifications solves our initial example.

Example 3.2.1. Let $\mathcal{D}$ be the ramification domain consisting of entity `pt`, fluent names $\texttt{runs}^0$ and $\texttt{in}^1$, and action name `ignite` accompanied by the action law `ignite` <u>transforms</u> $\{\neg\texttt{runs}\}$ <u>into</u> $\{\texttt{runs}\}$. Furthermore, let $\texttt{in}(x) \supset disq(\texttt{ignite})$ be a disqualifying condition. Suppose Σ be the transition model of $\mathcal{D}$, and let $\mathcal{O}_1$ consist of the single observation

$$\neg\texttt{runs} \ \underline{\text{after}}\ [\,]$$

Two models $M_1 = (\Sigma, Res_1, Ab_1)$ and $M_2 = (\Sigma, Res_2, Ab_2)$ exist for the scenario $(\mathcal{O}_1,\mathcal{D})$, namely, where $Res_1([\,]) = \{\neg\texttt{runs}, \neg\texttt{in(pt)}\}$ and $Res_2([\,]) = \{\neg\texttt{runs}, \texttt{in(pt)}\}$. Since the antecedent of the instance $\{x \mapsto \texttt{pt}\}$ of the underlying disqualifying condition, $\texttt{in}(x) \supset disq(\texttt{ignite})$, is true in $Res_2([\,])$, we have $Ab_2 = \{[\texttt{ignite}]\}$. In contrast, the first model entails no abnormal disqualification, that is, $Ab_1 = \{\}$. Hence, $M_1 \prec M_2$. Model M_1 being the unique preferred one, we see that our scenario entails the observation `runs` <u>after</u> [`ignite`].

Let, on the other hand, $\mathcal{O}_2$ consist of the two observations

$$\neg\texttt{runs} \ \underline{\text{after}}\ [\,]$$
$$\texttt{in(pt)} \ \underline{\text{after}}\ [\,]$$

then the only model for $(\mathcal{O}_2, \mathcal{D})$ is M_2 from above, which thus is preferred. The state $Res_2([\texttt{ignite}])$ is not defined due to our disqualifying condition. Hence, $(\mathcal{O}_2, \mathcal{D})$ entails the observation **ignite** inexecutable after [ ]. ■

This example demonstrates that our account of the Qualification Problem meets the basic requirement: Abnormal disqualifications may be assumed away by default, and in fact are so whenever this offers a coherent account of the observations. Consequently, the new entailment relation is nonmonotonic.

Our way of assuming away unlikely action disqualifications for the generation of preferred models can be characterized as globally minimizing abnormalities. No preference is given in case an action scenario implies that there be one out of two alternative abnormalities without necessitating a particular one. Suppose, for instance, observation suggests that there either be a potato in the tail pipe or a used chewing gum in the car's door lock thus rendering unexecutable the action of unlocking the door. This gives rise to two preferred models, one of which accounts for the first and the other for the second abnormal disqualification. The two models are considered equally plausible, which raises the question whether so doing might not be overly credulous. If, say, the neighborhood where we have left the car unattended for some time suffers from frequent strikes of a tail pipe marauder, then likelihood might favor the conclusion that we will be unable to start the engine while succeeding in unlocking the door.

One solution to this problem is the explicit introduction of probability values for each potential abnormal disqualification. In most real-world domains, however, it seems difficult if not impossible to acquire knowledge this precise. The Qualification Problem is therefore concerned with qualitative rather than quantitative reasoning about unlikely disqualifications of actions.[2] Accordingly, we will later (in Section 3.5) provide means to distinguish different degrees of (im-)probability. This will enable us to prefer minimization of abnormalities that are *a priori* more unlikely to occur than others.

The question remains whether global minimization of abnormal action disqualifications risks to ignore other fundamental reasons for preferring an abnormality over another one—that is, reasons which are not grounded on *a priori* differences in likelihood. Within an isolated state, preference can also be caused by observations entailing a certain abnormality. Our approach accounts for this via the notion of disqualifying conditions. On the other hand, the various states that are passed during the evolution of a system are not at all isolated. Consequently, preference of some abnormality over another one may also be caused by the dynamics of action theories. That is to say, a particular abnormal disqualification might naturally come about by state transition, in which case causality may provide good reasons for preferring this very abnormality over alternatives. The following simple extension of our introductory example illustrates this point and shows that our account

[2] More to nonmonotonic vs. probabilistic reasoning in the brief historical account in Section 3.7.

Figure 3.1. In general, we would consider it abnormal if we failed to start our car. But would we still do so if we deliberately insert a potato into the tail pipe beforehand?

of the Qualification Problem so far might fail to recognize intuitively expected preferences among abnormalities.

Example 3.2.2. Let $\mathcal{D}$ be the ramification domain of Example 3.2.1 but including the fluent name $\texttt{heavy}^1$, denoting whether its argument is a heavy object, and the action name $\texttt{insert}^1$, denoting the introduction of an object into the tail pipe. The new action is accompanied by the law

$$\texttt{insert}(x)\ \underline{\text{transforms}}\ \{\neg\texttt{in}(x)\}\ \underline{\text{into}}\ \{\texttt{in}(x)\}$$

The action is, however, abnormally disqualified if the object to be inserted is too heavy, which is accounted for by the disqualifying condition $\texttt{heavy}(x) \supset disq(\texttt{insert}(x))$ in addition to $\texttt{in}(x) \supset disq(\texttt{ignite})$.

Suppose it is known that the engine is not running and the tail pipe is not housing a potato. What, then, would be the expected outcome of first trying to insert a potato and, afterwards, trying to start the engine (c.f. Fig. 3.1)? Since there is no reason to believe that the potato is too heavy, the first action should be assumed qualified by default. The tail pipe thus becoming clogged disqualifies the second action. While this is the expected conclusion, our formal approach suggests equally plausible a second course of events. To see why, let $\mathcal{O}$ consist of the single observation $\neg\texttt{runs} \wedge \neg\texttt{in}(\texttt{pt})\ \underline{\text{after}}\ [\,]$. Suppose Σ is the transition model of $\mathcal{D}$, and let Res_1 and Res_2 be two partial mappings from action sequences to states designed as shown in Fig. 3.2. Then both interpretations $M_1 = (\Sigma, Res_1, Ab_1)$ with $Ab_1 = \{[\texttt{insert}(\texttt{pt}), \texttt{ignite}]\}$ and $M_2 = (\Sigma, Res_2, Ab_2)$ with $Ab_2 = \{[\texttt{insert}(\texttt{pt})]\})$ are models of $(\mathcal{O}, \mathcal{D})$. The first model, M_1, suggests that inserting the potato is successful and, hence, renders it impossible to start the engine afterwards—just as expected. Model M_2, on the other hand, declares the action of inserting the potato unqualified in the first place. The crucial point, now, is that by assuming this disqualification one avoids to grant a disqualification of starting the car. Consequently, $Ab_1 \not\subseteq Ab_2$, hence $M_1 \not\prec M_2$ (nor, of course, $M_2 \prec M_1$). Since there is obviously no model of $(\mathcal{O}, \mathcal{D})$ of the form $(\Sigma, Res, \{\})$,[3] both M_1 and M_2 are preferred. Thus the observation

[3] The reason being that either of $[\texttt{insert}(\texttt{pt})]$ or $[\texttt{insert}(\texttt{pt}), \texttt{ignite}]$ must be abnormally disqualified according to the underlying transition model in conjunction with the disqualifying conditions.

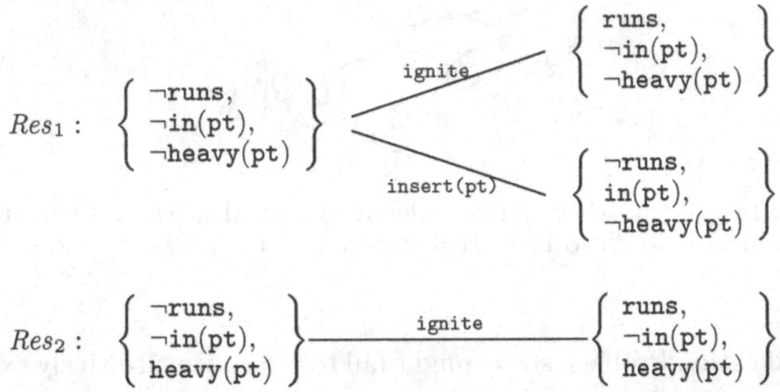

Figure 3.2. The scenario $\mathcal{O} = \{\neg\mathtt{runs} \wedge \neg\mathtt{in(pt)}\ \underline{\text{after}}\ []\}$ suggests two models. Our model preference criterion is not sophisticated enough to distinguish between them, although the one determined by Res_1 is much more plausible.

$$\mathtt{ignite}\ \underline{\text{inexecutable}}\ \underline{\text{after}}\ [\mathtt{insert(pt)}]$$

for instance, is not entailed, contrary to our expectations. ■

It should be stressed that the reason for the second model being counter-intuitive is not that in general potatoes being found in tail pipes is more likely than these being too heavy to lift. Rather we have encountered the general problem that global minimization of abnormalities does not allow to distinguish disqualifications which can be explained from the perspective of causality. Successfully introducing a potato into the tail pipe produces an effect which *causes* the fact that the second action, starting the engine, is unqualified. That is to say, while an abnormal disqualification of `insert(pt)` comes out of the blue in the unintended preferred model, M_2, an abnormal disqualification of `ignite`, as claimed in model M_1, is easily explicable. One even tends to not call the latter abnormal since being unable to start the engine after having clogged the tail pipe is, after all, what one can reasonably expect.

Thus we see that the dynamics of state transition possibly gives rise to a special kind of preference among abnormal disqualifications—a preference which is not entailed by observations nor by *a priori* differences in likelihood. Our account of the Qualification Problem therefore needs revision.

3.3 Causing Abnormal Action Disqualifications

When assuming away abnormal disqualifications of actions in order to tackle the Qualification Problem, care has to be taken that any such assumption is withheld in case specific information hints at the presence of abnormal circumstances. We have seen that hints of this kind may be revealed by the dynamics of state transition, a fact which is ignored if the preference criterion is given by global minimization. Abnormalities caused by the performance of actions need to be preferably accepted or, for that matter, not considered abnormal at all. The action of starting the car engine being disqualified, for instance, is a perfectly reasonable, since natural, consequence of introducing a potato into the tail pipe.

With the last remark we seem to have pinned down the problem to the question of how to account for the fact that abnormalities may be obtained as a side effect of other actions. That is to say, the occurrence of an action disqualification is not abnormal if being evoked as indirect effect of another action. Put that way, the whole apparatus of Chapter 2 might furnish a ready fundamental for a solution to the problem of caused abnormalities: Suppose the expression $disq(a)$ is a *fluent* stating whether or not action a is abnormally disqualified in a state. In the light of this new interpretation, a disqualifying condition $F \supset disq(a)$ reveals itself as state constraint so that $disq(a)$ holds whenever F is true. In particular, any such constraint should give rise to the indirect effect $disq(a)$ whenever F is brought about. Fluent $\mathtt{in(pt)}$ being made true by some action, for instance, is thus expected to trigger the indirect effect $disq(\mathtt{ignite})$ according to $\mathtt{in}(x) \supset disq(\mathtt{ignite})$, now to be taken as state constraint.

Fluents representing disqualifications in conjunction with suitable state constraints enable us to model, by means of ramification, situations where an abnormality is caused by other actions. Whenever some $disq(a)$ is initiated in this way, then that should no longer receive special attention insofar as comparing the abnormalities occurring in models is concerned. This reflects the intention to consider normal any case of abnormal action disqualifications which are deliberately brought about. The question remains, however, how the 'real' abnormalities are assumed away by default. If, as we have argued, this cannot be done globally over all states, then what is the alternative?

The answer is that abnormalities are to be minimized solely in the initial state and so to leave it to both persistence and ramification to take care of the correct evolution of these fluents as time advances. We recall Example 3.2.2 to supply the reader with a feeling of how that strategy solves the problem of causally motivated preferences among abnormalities. Let the two fluent formulas

$$\begin{array}{l} \exists x.\, \mathtt{in}(x) \supset disq(\mathtt{ignite}) \\ \forall x\, [\, \mathtt{heavy}(x) \supset disq(\mathtt{insert}(x))\,] \end{array} \tag{3.1}$$

be state constraints. Given the observation $\neg\mathtt{runs} \wedge \neg\mathtt{in(pt)}$ <u>after</u> $[\,]$, it is consistent to assume the initial state be

$$S = \{\neg\mathtt{runs}, \neg\mathtt{in(pt)}, \neg\mathtt{heavy(pt)}, \neg disq(\mathtt{ignite}), \neg disq(\mathtt{insert(pt)})\,\}$$

Noticeably, all abnormalities are denied in this state. It is clear, therefore, that S is preferred according to the proposed criterion. As a matter of fact, it is the only state that possesses this property. For if $\mathtt{heavy(pt)}$ were true, then this would entail an abnormality according to the second state constraint of (3.1). Having both $\neg disq(\texttt{insert(pt)}) \in S$ and $\neg\texttt{in(pt)} \in S$ (the only regular precondition of inserting a potato), it follows that action insert(pt) is executable in S. The direct effect being in(pt), we expect the additional, indirect effect $disq(\texttt{ignite})$ deriving from the first one of our state constraints (3.1). The successor state resulting from performing insert(pt) thus correctly entails an abnormal disqualification of action ignite. To stress the point, we again draw the reader's attention to the fact that the initial state S, although entailing an abnormality at a later timepoint, is in itself free of abnormalities.

Restricting minimization of abnormalities to the initial state thus provides a solution to our key example that proved global minimization inadequate as a formal account of the Qualification Problem. The question immediately rises, then, whether the improved approach is generally well-founded and not refutable itself. To begin with, the new minimization strategy is bound to fail in case it neglects preferences apparently to be made among alternative ways of minimizing in specific situations. Now, any such preference either grounds on static reasons, that is, it holds in any state regardless of both past and future states, or the reasons for a preference lie in the dynamics of state evolution. As regards static preferences, different degrees of *a priori* likelihood of abnormalities shall be neglected for the moment and dealt with in Section 3.5. For the moment the only possible static reason for preferring to accept an abnormality is additional information implying its presence. To account for this is the role of state constraints defining conditions for a fluent $disq(a)$ being true. Since any acceptable state must satisfy these constraints by definition, our approach does respect static reasons for preferring to accept an abnormality.

Reasons to accept an abnormality grounded on the dynamics of state transition are linked to causality. That is to say, the acceptance of an abnormal disqualification may be founded on the performance of certain actions. Let us take for granted the unidirectionality of causality forward in time.[4] Two points need to be made then. First, suppose that at some stage an action is performed which brings about a condition for an action a being abnormally disqualified. Whenever this happens, the corresponding fluent $disq(a)$ should become true as side effect of the other action. This is achieved by any state constraint of the form $F \supset disq(a)$ giving rise to the indirect effect $disq(a)$ as soon as F gets invoked. Our solution to the Ramification Problem guarantees this—provided that the underlying influence information entails that any fluent occurring in F potentially affects $disq(a)$.[5]

[4] This is not meant as offense to the minority of philosophers who take serious the idea of effects preceding their causes, nor to the minority of physicists who interpret quantum mechanics in such a way that it gives rise to backward causality. We appeal to common sense here.

[5] This holds even if one abandons the rather idealistic assumption that a domain specification contain complete knowledge as to the possible reasons for an action disqualification. The trick is to introduce an artificial fluent representing the 'unknown' cause; see Section 3.5 below.

The second point to be made, which completes the overall argument, is to justify our minimizing abnormalities initially. But given that causality is effective only forward in time, it is clear that no causal reason for an abnormality in the initial state can possibly be known of. This by no means implies that such a causal reason does not exist. But if it does, then it must lie outside the scenario specification, hence has no influence on the correct reasoning about this scenario.

Having justified our strategy of how to cope with the Qualification Problem, we proceed with developing a formal account of this approach. To begin with, any so-called qualification domain is supposed to include, for each action a, the fluent $disq(a)$ stating whether or not action a is abnormally disqualified in a state. Abnormal disqualifications indicate abnormal circumstances. The latter might be specified with the help of fluents which, too, are expected to be assumed false by default. Example fluents of this kind might be `in(pt)` and `heavy(pt)`, as normally tail pipes are not clogged by potatoes, let alone the possibility of a potato being too heavy to lift. Fluents describing abnormal circumstances can be combined in state constraints to describe the conditions for a particular action being abnormally disqualified. We make no formal presuppositions as to the structure of these constraints, but we will later, in the following section, argue for a general strategy of how to design them.

Definition 3.3.1. *A* plain[6] qualification domain $\mathcal{D}$ *is a ramification domain with a distinguished subset* $\mathcal{F}_{ab}$, *called* abnormality fluents, *of the set of all fluents so that* $disq(a) \in \mathcal{F}_{ab}$ *for each action* a. *The* transition model *of* $\mathcal{D}$ *is a mapping* Σ *from pairs of an acceptable state and an action into (possibly empty) sets of states such that* $S' \in \Sigma(S, a)$ *iff* $\neg disq(a) \in S$ *and* S' *is successor of* S *and* a. ■

Notice that the transition model of a qualification domain assigns no next state whenever an action is abnormally disqualified, regardless of whether successor states exist.

Like in the approach based on global minimization, the new minimization policy is realized by means of a model preference criterion. Informally speaking, the less abnormality fluents are initially true the better.

Definition 3.3.2. *A* plain qualification scenario *is a pair* $(\mathcal{O}, \mathcal{D})$ *where* $\mathcal{D}$ *is a plain qualification domain and* $\mathcal{O}$ *is a set of observations. An* interpretation *for* $(\mathcal{O}, \mathcal{D})$ *is a pair* (Σ, Res) *where* Σ *is the transition model of* $\mathcal{D}$ *and* Res *is a partial function which maps finite (possibly empty) action sequences to acceptable states and which satisfies the following:*

1. $Res([\,])$ *is defined.*
2. *For any sequence* $a^* = [a_1, \ldots, a_{k-1}, a_k]$ *of actions* $(k > 0)$,

[6] The reason for calling "plain" these qualification domains is that for the moment all abnormalities are considered equally likely.

a) $Res(a^*)$ is defined if and only if $Res([a_1, \dots, a_{k-1}])$ is defined and $\Sigma(Res([a_1, \dots, a_{k-1}]), a_k)$ is not empty, and
b) $Res(a^*) \in \Sigma(Res([a_1, \dots, a_{k-1}]), a_k)$.

If $I = (\Sigma, Res)$ and $I' = (\Sigma, Res')$ are interpretations, then I is less abnormal *than I', written $I \prec I'$, iff $Res([\,]) \cap \mathcal{F}_{ab} \subsetneq Res'([\,]) \cap \mathcal{F}_{ab}$. A* model *of a plain qualification scenario $(\mathcal{O}, \mathcal{D})$ is an interpretation (Σ, Res) such that each $o \in \mathcal{O}$ is true in Res. A model is* preferred *iff there is no other model which is less abnormal. An observation o is* entailed *iff it is true in all preferred models of $(\mathcal{O}, \mathcal{D})$.* ■

Example 3.3.1. Let $\mathcal{D}$ be the ramification domain of Example 3.2.2 but now including the fluent names $disq(\texttt{ignite})^0$ and $disq(\texttt{insert})^1$ plus the steady state constraints

$$\exists x.\, \texttt{in}(x) \supset disq(\texttt{ignite})$$
$$\forall x\, [\, \texttt{heavy}(x) \supset disq(\texttt{insert}(x))\,]$$

Consider $\mathcal{F}_{ab} = \{\texttt{in(pt)}, \texttt{heavy(pt)}, disq(\texttt{ignite}), disq(\texttt{insert(pt)})\}$, then $\mathcal{D}$ is a (plain) qualification domain. Let $\mathcal{O}$ consist of the observation

$$\neg\texttt{runs}\ \underline{\text{after}}\ [\,]$$

then $(\mathcal{O}, \mathcal{D})$ is a qualification scenario. Suppose Σ be the transition model of $\mathcal{D}$. The domain being deterministic, interpretations (Σ, Res) are uniquely characterized by the initial state $Res([\,])$, e.g.

$$Res_1([\,]) = \{\neg\texttt{runs}, \neg\texttt{in(pt)}, \neg\texttt{heavy(pt)}, \neg disq(\texttt{ignite}), \neg disq(\texttt{insert(pt)})\}$$

Obviously, (Σ, Res_1) is a model of $(\mathcal{O}, \mathcal{D})$ as it satisfies $\neg\texttt{runs} \in Res_1([\,])$. Since $Res_1([\,]) \cap \mathcal{F}_{ab} = \{\}$, this model is preferred. Moreover, it is unique in that respect because the initial value of the only existing non-abnormality fluent, viz. `runs`, is fixed by the given observation. Thus whatever is true in this model is also entailed by our scenario, $(\mathcal{O}, \mathcal{D})$. In particular, we have $\texttt{in(pt)} \in Res_1([\texttt{insert(pt)}])$ and, hence, $disq(\texttt{ignite}) \in Res_1([\texttt{insert(pt)}])$ according to transition model Σ. Consequently, $(\mathcal{O}, \mathcal{D})$ entails both the two observations

$$\texttt{in(pt)}\ \underline{\text{after}}\ [\texttt{insert(pt)}]$$
$$\texttt{ignite}\ \underline{\text{inexecutable}}\ \underline{\text{after}}\ [\texttt{insert(pt)}]$$

■

3.4 Conditions for Abnormal Action Disqualifications

Up to this point our discussion of the Qualification Problem revolved around the questions both of how to assume away abnormal disqualifications and of how to withdraw this assumption in case some condition for the corresponding abnormality is entailed. A related task is of equal importance for reasoning about actions in environments which are unpredictable insofar as success of actions cannot be guaranteed. Suppose an action turns out unexecutable. This should come as a surprise whenever all regular preconditions were satisfied and there was no reason to believe in an abnormality. Having nonetheless encountered the disqualification, it is both natural and reasonable to seek an explanation. An autonomous agent, for instance, whose current goal relies on the successful performance of the action that failed should try to figure out what went wrong in order to rectify it, if possible. This raises the question of what qualifies as an adequate explanation for an abnormality.

The natural thing to do if an abnormal disqualification has been observed is to search the available domain knowledge for conditions which would entail it. Among all potential causes those offer an explanation which are compatible with the entire state of affairs. Suppose, for instance, our protagonist knows that starting the car is impossible if some object clogs the tail pipe, if the tank is empty, the battery is low, or if there is a general problem with the engine itself. Suppose further that she has checked both tail pipe and tank and has also confirmed the good status of the battery, say, by turning on the radio. To the best of her knowledge, then, failing to start the car must be caused by an engine problem.

On the formal side, this generation of explanations for observed abnormalities requires a specific policy of designing the state constraints that relate action disqualifications to their conditions. To see why, suppose the following four constraints specify all that is known as to reasons for failing to start a car.

$$\begin{aligned} \exists x.\, \mathtt{in}(x) &\supset \mathit{disq}(\mathtt{ignite}) \\ \mathtt{tank\text{-}empty} &\supset \mathit{disq}(\mathtt{ignite}) \\ \mathtt{low\text{-}battery} &\supset \mathit{disq}(\mathtt{ignite}) \\ \mathtt{engine\text{-}problem} &\supset \mathit{disq}(\mathtt{ignite}) \end{aligned} \tag{3.2}$$

The conjunction of these formulas is logically equivalent to

$$\begin{aligned} \exists x.\, \mathtt{in}(x) \vee \mathtt{tank\text{-}empty} \vee \mathtt{low\text{-}battery} \vee \mathtt{engine\text{-}problem} \\ \supset \mathit{disq}(\mathtt{ignite}) \end{aligned} \tag{3.3}$$

This being a fluent formula that any acceptable state must satisfy, whenever at least one of the four (abnormal) causes for a disqualification of `ignite` holds at some point, then $\mathit{disq}(\mathtt{ignite})$ holds as well. The converse, however, is not necessarily true. That is to say, $\mathit{disq}(\mathtt{ignite})$ may hold in a state without any of the potential causes on the left hand side of the above implication being true. Even worse, the four conditions supposedly belong to the set of abnormality fluents since each describes rather unusual circumstances. Consequently, not only do models exist which entail no explanation for $disq(\texttt{ignite})$ in case the latter is known to be true, these models are even preferred.

The desired proliferation of explanations can, however, be achieved by a modification known as *completion* of the fluent formula (3.3). The problem with this implication, as it stands, being that it does not support any further conclusions from $disq(\texttt{ignite})$, it is transformed into a bi-conditional:

$$\begin{aligned} \exists x.\, \texttt{in}(x) \vee \texttt{tank-empty} \vee \texttt{low-battery} \vee \texttt{engine-problem} \\ \equiv disq(\texttt{ignite}) \end{aligned} \tag{3.4}$$

Let this formula replace the state constraints listed in (3.2). Then any acceptable state which satisfies $disq(\texttt{ignite})$ must also satisfy at least one of $\exists x.\, \texttt{in}(x)$, **tank-empty**, **low-battery**, or **engine-problem**. The three observations

$$\neg\texttt{runs}\ \underline{\text{after}}\ [\,]$$

$$\texttt{ignite}\ \underline{\text{inexecutable}}\ \underline{\text{after}}\ [\,]$$

$$\forall x.\, \neg\texttt{in}(x) \wedge \neg\texttt{tank-empty} \wedge \neg\texttt{low-battery}\ \underline{\text{after}}\ [\,]$$

for instance, entail **engine-problem** <u>after</u> [] on the basis of the completed state constraint.

The mechanism of completion provides us with a general strategy of how to design state constraints so as to automatically supply explanations for observed action disqualifications. Suppose n different abnormal conditions, specified by fluent formulas $F_1, \ldots, F_n$, are known to render impossible the execution of action a. Instead of introducing the n implicational state constraints $F_1 \supset disq(a), \ldots, F_n \supset disq(a)$, we employ a (stronger) bi-conditional as follows.

$$\bigvee_{i=1}^{n} F_i \;\equiv\; disq(a) \tag{3.5}$$

It obviously entails the aforementioned n implications, but in addition allows to reason the other direction, i.e., from abnormalities to their possible causes. While this is an invaluable gain of using completion, there is, however, an important objection against state constraints of the form (3.5). Namely, equating a disqualification with a disjunction of conditions presupposes complete knowledge as to the possible reasons for an action to fail. No doubt, abnormal action disqualifications being rare exceptions by definition, the situation is even more exceptional where this disqualification cannot be explained. Yet, not accounting for these cases seems to ignore the fact that generally only partial knowledge can be acquired of real-world domains. Completion being too valuable to abandon, however, we favor the development of additional means to deal with inexplicable abnormalities.

Prior to so doing, let us mention another crucial issue when dealing with action disqualifications, which stresses the importance of completed state constraints. Suppose our protagonist clears the tail pipe again after having successfully introduced a potato. If initially starting the engine has been assumed qualified by default, then this assumption had to be withdrawn after having clogged the tail pipe. But the assumption of qualification is to be re-installed in the following situation where the potato has been removed again. For if prior to performing the two actions nothing hints at an abnormality, then there is no reason to believe that things have been changed by these two actions with reverse effect. This revoking qualification is granted if state constraints of the form (3.5) are employed. To see why, suppose that initially each F_i $(1 \leq i \leq n)$ and, hence, $disq(a)$ is assumed false by default. Suppose further that later on one particular $F_k \in \{F_1, \ldots, F_n\}$ is initiated and that sometime thereafter F_k is terminated again. Then $\neg disq(a)$ must hold at the end. This follows from persistence of each $\neg F_j$ $(1 \leq j \leq n,\ j \neq k)$ and from the fact that $\bigwedge_{i=1}^{n} \neg F_i$ implies $\neg disq(a)$ according to the completed state constraint.

Let us summarize and illustrate the discussion on conditions for abnormal action disqualifications by a fully formalized example which highlights the essential aspects.

Example 3.4.1. Let $\mathcal{D}$ be the plain qualification domain consisting of the entity pt, the fluent names $\mathcal{F} = \{\texttt{runs}^0, \texttt{in}^1, \texttt{tank-empty}^0, \texttt{low-battery}^0, \texttt{engine-problem}^0, \texttt{heavy}^1, disq(\texttt{ignite})^0, disq(\texttt{insert})^1, disq(\texttt{clear})^0\}$ such that only runs is a non-abnormality fluent, the action names $\mathcal{A} = \{\texttt{ignite}^0, \texttt{insert}^1, \texttt{clear}^0\}$, the action laws

$$\begin{array}{rlll} \texttt{ignite} & \underline{\text{transforms}} & \{\neg\texttt{runs}\} & \underline{\text{into}} \ \{\texttt{runs}\} \\ \texttt{insert}(x) & \underline{\text{transforms}} & \{\neg\texttt{in}(x)\} & \underline{\text{into}} \ \{\texttt{in}(x)\} \\ \texttt{clear} & \underline{\text{transforms}} & \{\texttt{in}(x)\} & \underline{\text{into}} \ \{\neg\texttt{in}(x)\} \end{array}$$

the three steady state constraints

$$\begin{array}{rl} [\exists x.\, \texttt{in}(x) \vee \texttt{tank-empty} \vee \texttt{low-battery} \vee \texttt{engine-problem} & \\ \equiv disq(\texttt{ignite})] & \\ \texttt{heavy}(x) \equiv disq(\texttt{insert}(x)) & \\ \bot \equiv disq(\texttt{clear}) & \end{array}$$

and, obtainable through the influence information

$$\begin{array}{l} \mathcal{I} = \{(\texttt{heavy}(x), disq(\texttt{insert}(x)))\} \\ \quad \cup \{(f, disq(\texttt{ignite})) : \\ \qquad f \in \{\texttt{in}(x), \texttt{tank-empty}, \texttt{low-battery}, \texttt{engine-problem}\}\} \end{array}$$

these ten (steady) causal relationships:

$$
\begin{array}{rlll}
\texttt{in}(x) & \underline{\text{causes}} & \mathit{disq}(\texttt{ignite}) & \underline{\text{if}}\ \top \\
\texttt{tank-empty} & \underline{\text{causes}} & \mathit{disq}(\texttt{ignite}) & \underline{\text{if}}\ \top \\
\texttt{low-battery} & \underline{\text{causes}} & \mathit{disq}(\texttt{ignite}) & \underline{\text{if}}\ \top \\
\texttt{engine-problem} & \underline{\text{causes}} & \mathit{disq}(\texttt{ignite}) & \underline{\text{if}}\ \top \\
\neg\texttt{in}(x) & \underline{\text{causes}} & \neg\mathit{disq}(\texttt{ignite}) & \\
 & \underline{\text{if}} & \forall y.\, \neg\texttt{in}(y) \wedge \neg\texttt{tank-empty} & \\
 & & \wedge \neg\texttt{low-battery} \wedge \neg\texttt{engine-problem} & \\
\neg\texttt{tank-empty} & \underline{\text{causes}} & \neg\mathit{disq}(\texttt{ignite}) & \\
 & \underline{\text{if}} & \forall y.\, \neg\texttt{in}(y) \wedge \neg\texttt{low-battery} & \\
 & & \wedge \neg\texttt{engine-problem} & \\
\neg\texttt{low-battery} & \underline{\text{causes}} & \neg\mathit{disq}(\texttt{ignite}) & \\
 & \underline{\text{if}} & \forall y.\, \neg\texttt{in}(y) \wedge \neg\texttt{tank-empty} & \\
 & & \wedge \neg\texttt{engine-problem} & \\
\neg\texttt{engine-problem} & \underline{\text{causes}} & \neg\mathit{disq}(\texttt{ignite}) & \\
 & \underline{\text{if}} & \forall y.\, \neg\texttt{in}(y) \wedge \neg\texttt{tank-empty} \wedge \neg\texttt{low-battery} & \\
\texttt{heavy}(x) & \underline{\text{causes}} & \mathit{disq}(\texttt{insert}(x)) & \underline{\text{if}}\ \top \\
\neg\texttt{heavy}(x) & \underline{\text{causes}} & \neg\mathit{disq}(\texttt{insert}(x)) & \underline{\text{if}}\ \top
\end{array}
$$

1. Let $\mathcal{O}_1$ consist of the observations

$$\neg\texttt{runs}\ \underline{\text{after}}\ [\,]$$

$$\texttt{ignite}\ \underline{\text{inexecutable}}\ \underline{\text{after}}\ [\,]$$

$$\forall x.\, \neg\texttt{in}(x) \wedge \neg\texttt{tank-empty} \wedge \neg\texttt{low-battery}\ \underline{\text{after}}\ [\,]$$

Then the qualification scenario $(\mathcal{O}_1, \mathcal{D})$ entails $\texttt{engine-problem}\ \underline{\text{after}}\ [\,]$.

2. Let $\mathcal{O}_2$ consist of the only observation

$$\texttt{tank-empty}\ \underline{\text{after}}\ [\,]$$

Then the scenario $(\mathcal{O}_2, \mathcal{D})$ entails $\texttt{ignite}\ \underline{\text{inexecutable}}\ \underline{\text{after}}\ [\,]$. Notice that despite this disqualification any preferred model (Σ, Res) of the scenario satisfies $\neg\texttt{in}(\texttt{pt}), \neg\texttt{low-battery}, \neg\texttt{engine-problem} \in \mathit{Res}([\,])$, for all these fluents belong to the set of abnormality fluents. Consequently, $(\mathcal{O}_2, \mathcal{D})$ also entails

$$\forall x.\, \neg\texttt{in}(x) \wedge \neg\texttt{low-battery}, \neg\texttt{engine-problem}\ \underline{\text{after}}\ [\,]$$

That is to say, although action `ignite` is known to be abnormally disqualified, we still conclude, by default, that the tail pipe is clear and that both battery and engine are in order.

3. Finally, let $\mathcal{O}_3$ consist of the observation

$$\neg\texttt{runs}\ \underline{\text{after}}\ [\,]$$

Then the scenario $(\mathcal{O}_3, \mathcal{D})$ entails $\texttt{in(pt)}$ <u>after</u> $[\texttt{insert(pt)}]$, hence also $\texttt{ignite}$ <u>inexecutable</u> <u>after</u> $[\texttt{insert(pt)}]$. Moreover, the observation $\neg\texttt{in(pt)}$ <u>after</u> $[\texttt{insert(pt)}, \texttt{clear}]$ and, hence, the observation $\neg \mathit{disq}(\texttt{ignite})$ <u>after</u> $[\texttt{insert(pt)}, \texttt{clear}]$ are entailed. Mainly responsible for this last conclusion is the causal relationship

$$\neg\texttt{in}(x) \ \underline{\text{causes}} \ \neg \mathit{disq}(\texttt{ignite}) \ \underline{\text{if}} \ \forall y. \neg\texttt{in}(y) \land \neg\texttt{tank-empty} \\ \land \neg\texttt{low-battery} \\ \land \neg\texttt{engine-problem}$$

Thus, in addition to the above, $(\mathcal{O}_3, \mathcal{D})$ entails

$$\texttt{runs} \ \underline{\text{after}} \ [\texttt{insert(pt)}, \texttt{clear}, \texttt{ignite}]$$

which shows how qualification gets revoked once the only reason for a disqualification disappears.

■

3.5 Degrees of Abnormality

Our account of the Qualification Problem so far makes two simplifying assumptions. First, all abnormalities are considered equally unlikely. Second, complete knowledge as to all possible causes for an abnormal disqualification is implicitly assumed by equating the abnormality with these causes. A natural extension of the current theory is to allow different degrees of abnormality. In this way, it becomes possible to specify, for instance, that having been running out of gas is more likely than a low battery, which in turn is more plausible an explanation for being unable to start the car than a potato being placed in the tail pipe. Different levels of likelihood are thus representable without the necessity to provide precise probabilities.

The introduction of degrees of abnormality incidentally also provides a ready means to resolves the problem associated with our equating abnormal disqualifications with their possible causes. In addition to all conceivable reasons for an abnormality to occur, one fluent in the list of causes may represent the *unknown* cause. Suppose a domain constraint $\mathit{disq}(a) \equiv \bigvee_i F_i$ is weakened to the effect that $\mathit{disq}(a) \equiv \bigvee_i F_i \lor \texttt{mysterious}(a)$. Then it is no longer violated in case a has been observed abnormally disqualified while each known cause, F_i, has been proved false.[7] Suppose further that 'abnormality' fluent $\texttt{mysterious}(a)$ has higher degree of improbability than

[7] Introducing the unknown cause allows to model, for example, the tendency of some human beings to hit out at mechanical devices, e.g., dispensing machines, when discovering their malfunctioning. The foregoing reasoning process there, if any, attributes the encountered malfunction to some unobservable fluent which is hoped to being manipulable by means of that very action.

any other, concrete cause. Then the unknown cause is used only as a last resort when it comes to explaining an abnormality.

On the formal side, degrees of abnormality are represented by a partial ordering among the 'abnormality' fluents. If, for instance, we specify that **tank-empty** $<$ **low-battery**, then **low-battery** shall always be minimized—initially—with higher priority than **tank-empty**. Accordingly, the concept of unknown causes is adequately modeled by defining $f_{ab} < \mathtt{mysterious}(a)$ for all other 'abnormality' fluents f_{ab}. Property $\mathtt{mysterious}(a)$ is thus assigned the highest degree of abnormality. Being a partial ordering, the comparison relation $<$ may be indifferent regarding some pairs of fluents, in which case no preference is made for either of them. For a formal introduction to partial orderings and related concepts see Annotation 3.2. In the following, we extend our solution to the Qualification Problem to the effect that different degrees of abnormality are supported.

A *binary relation* R on some set A is a subset of the Cartesian product $A \times A$, that is, R is a set of pairs (a, b) where $a, b \in A$. If $(a, b) \in R$, then this is also written aRb. A binary relation R may obey the following properties:

$$\begin{array}{rll} \forall a \in A. & \neg\, aRa & \text{(irreflexive)} \\ \forall a, b \in A. & (aRb \wedge bRa \supset a = b) & \text{(antisymmetric)} \\ \forall a, b, c \in A. & (aRb \wedge bRc \supset aRc) & \text{(transitive)} \end{array}$$

If it does, then the relation R is a *partial ordering* on A. If in addition R satisfies $\forall a, b \in A.\ (aRb \vee bRa)$, then R is *strict*. A strict ordering R' is an *extension* of a partial ordering R iff $R' \supseteq R$, that is, whenever aRb then also $aR'b$. Let, for example, A be

$$\{\mathtt{tank\text{-}empty}, \mathtt{low\text{-}battery}, \mathtt{engine\text{-}problem}, \mathtt{in(pt)}, \mathtt{mysterious(ignite)}\}$$

and suppose a partial ordering $<$ be given by

$$\begin{array}{l} \mathtt{tank\text{-}empty} < \mathtt{low\text{-}battery} < \mathtt{in(pt)} < \mathtt{mysterious(ignite)} \\ \mathtt{tank\text{-}empty} < \mathtt{engine\text{-}problem} < \mathtt{mysterious(ignite)} \end{array}$$

Let $\ll$ be $<$ augmented by $\mathtt{low\text{-}battery} \ll \mathtt{in(pt)} \ll \mathtt{engine\text{-}problem}$, then $\ll$ is one out of three possible strict orderings extending $<$.

Annotation 3.2. Orderings.

Definition 3.5.1. *A* qualification domain $\mathcal{D}$ *is a plain qualification domain augmented by a partial ordering* $<$ *on the set of abnormality fluents. Accordingly, a* qualification scenario $(\mathcal{O}, \mathcal{D})$ *consists of a set* $\mathcal{O}$ *of observations and a qualification domain* $\mathcal{D}$.

Suppose $M = (\Sigma, Res)$ *is a model of* $(\mathcal{O}, \mathcal{D})$. *Then* M *is* preferred *iff there is a strict ordering* $\ll$ *which extends* $<$ *and such that for all mod-*

els $M' = (\Sigma, Res')$ *of* $(\mathcal{O}, \mathcal{D})$ *and all* $f_{ab} \in \mathcal{F}_{ab}$*: If* $f_{ab} \in Res([\,]) \setminus Res'([\,])$*, then there is some* $f_{ab}{}' \ll f_{ab}$ *such that* $f_{ab}{}' \in Res'([\,]) \setminus Res([\,])$. ■

In words, a preferred model is obtained by first choosing a minimization strategy, that is, a strict ordering which respects the given partial one. Then the model is preferred whose evolution function *Res* satisfies the following: Suppose some abnormality fluent f_{ab} is initially true in *Res* but false in the evolution function *Res'* of some other model. Then there must be another abnormality fluent $f_{ab}{}'$ of higher priority than f_{ab} according to the chosen strict ordering and which is initially false in *Res* but true in *Res'*.

Example 3.5.1. Let $\mathcal{D}$ be the qualification domain of Example 3.4.1 but with abnormality fluent `mysterious(ignite)` added as an additional cause for *disq*(`ignite`). Furthermore, let the following information regarding degrees of abnormality be given:

`tank-empty < low-battery < in(pt) < mysterious(ignite)`
`tank-empty < engine-problem < mysterious(ignite)`

Suppose Σ be the transition model of $\mathcal{D}$, and let $\mathcal{O}$ consist of the observations

$\neg$`runs` <u>after</u> [ ]
`ignite` <u>inexecutable</u> <u>after</u> [ ]
$\neg$`tank-empty` <u>after</u> [ ]

Given that `tank-empty` is false initially in any model of $(\mathcal{O}, \mathcal{D})$, each of `low-battery`, `engine-problem`, `in(pt)`, `mysterious(ignite)` offers as explanation for the observed disqualification. Following the *a priori* knowledge of likelihood given by $<$, we obtain two preferred models, namely, $M_1 = (\Sigma, Res_1)$, where $Res_1([\,]) \cap \mathcal{F}_{ab} = \{$`low-battery`, *disq*(`ignite`)$\}$, and $M_2 = (\Sigma, Res_2)$, where $Res_2([\,]) \cap \mathcal{F}_{ab} = \{$`engine-problem`, *disq*(`ignite`)$\}$.[8] Consequently, `low-battery` $\vee$ `engine-problem` <u>after</u> [ ] is entailed by the qualification domain $(\mathcal{O}, \mathcal{D})$, as is $\neg$`in(pt)` $\wedge$ $\neg$`mysterious(ignite)` <u>after</u> [ ]. Highly unlikely causes such as a potato being in the tail pipe, let alone a mysterious disqualification, are thus assumed away. ■

This completes our formal account of the Qualification Problem. Let us summarize: A qualification domain is supposed to contain a distinguished set of fluents $\mathcal{F}_{ab}$, each of which describes abnormal circumstances and thus is to be assumed false by default. This assumption, however, needs to be restricted to the initial state, so that these fluents are subject to the general law of persistence but are also potentially (directly or indirectly) affected by the performance of actions. Among these so-called abnormality fluents are expressions, denoted *disq*(a), which represent the property of action a being abnormally disqualified. State constraints relating *disq*(a) with possible

[8] Tacitly assuming that both Res_1 and Res_2 are otherwise arbitrary but satisfy the conditions of Definition 3.3.2.

causes support the proliferation of explanations in case an abnormal disqualification surprisingly occurs. Additional information may be provided as to different degrees of abnormality. The latter also furnishes a ready approach to accommodate situations in which a suitable explanation is not possible. The default assumption of 'normality' is formally captured by a model preference criterion, which induces a nonmonotonic entailment relation—a feature inherently connected with the Qualification Problem.

3.6 A Fluent Calculus Axiomatization

The adaptation of our Fluent Calculus-based axiomatization at the end of Chapter 2 to qualification domains and scenarios requires two issues be addressed. First, the notion of action sequences being qualified needs refinement. The existence of a successor state is still necessary but no longer sufficient a criterion. What needs additionally be guaranteed is that no abnormal disqualification is present. Modifying the axiomatization to this extent is rather straightforward. Second, and more substantial, some qualitatively new mechanism needs to be introduced to account for nonmonotonicity as an intrinsic feature of the Qualification Problem. Recall that our axiomatization of ramification domains and scenarios uses pure classical logic, hence is monotonic. The nonmonotonic framework in which it is to be embedded serves the purpose of assuming away abnormality by default.

3.6.1 Axiomatizing Qualification

The concept of action sequences being qualified has been defined, in the axiomatization of ramification in Chapter 2, by the following two axioms.

$$\begin{gathered} \mathit{Qualified}([\,]) \\ \mathit{Qualified}([a^*|a]) \equiv \mathit{Qualified}(a^*) \wedge \exists s, s' \begin{bmatrix} \mathit{Result}(a^*, s) \wedge \\ \mathit{Successor}(s, a, s') \end{bmatrix} \end{gathered} \tag{3.6}$$

(The reader may recall that $\mathit{Qualified}(a^*)$ indicates that action sequence a^* is qualified, i.e., executable; $\mathit{Result}(a^*, s)$ determines state s as the result of performing a^*; and $\mathit{Successor}(s, a, s')$ is true iff s' is a successor state of s and action a.) These axioms reflect the (idealistic) view that an action can be successfully performed whenever the regular preconditions are met. When addressing the Qualification Problem, an additional condition needs to be accounted for: The action in question must not be abnormally disqualified. This can be axiomatized with the help of the fluents $\mathit{disq}(a)$. The new definition of qualification, replacing (3.6), is as follows:

$$\mathit{Qualified}([\,]) \tag{3.7}$$

$$Qualified([a^*|a]) \equiv Qualified(a^*) \wedge \exists s, s' \begin{bmatrix} Result(a^*, s) \wedge \\ Holds(\neg disq(a), s) \wedge \\ Successor(s, a, s') \end{bmatrix} \tag{3.8}$$

The second axiom defines, in words, an action sequence $[a^*|a]$ to be qualified if so is a^*, if the result s of performing a^* does not entail an abnormal disqualification as regards a, and if there exists a successor s' of s and a. The reader may liken this to the notion of the model component *Res* being defined for the argument $[a^*|a]$ (c.f. Definitions 3.3.1 and 3.3.2 in Section 3.3). No further foundational axioms introduced in Chapter 2 require replacement.

Example 3.6.1. Let $\mathcal{D}$ be the qualification domain consisting of the entity $\texttt{pt}$; fluent names $\texttt{runs}^0$, $\texttt{in}^1$, $\texttt{mysterious(ignite)}^0$, $disq(\texttt{ignite})^0$, so that only $\texttt{runs}$ is an abnormality fluent; the partial priority ordering given by $\texttt{in(pt)} < \texttt{mysterious(ignite)}$; action name $\texttt{ignite}^0$; action law $\texttt{ignite}$ <u>transforms</u> $\{\neg\texttt{runs}\}$ <u>into</u> $\{\texttt{runs}\}$; and the steady state constraint

$$disq(\texttt{ignite}) \equiv \exists x.\, \texttt{in}(x) \vee \texttt{mysterious(ignite)}$$

Let $\mathcal{FC}_\mathcal{D}$ denote the Fluent Calculus-based axiomatization of this domain as described in Chapter 2 (c.f. Section 2.9.2). Consider, then, the collection of classical formulas $\mathcal{FC}_\mathcal{D} \cup \{(3.7),(3.8),(2.37)\text{–}(2.40)\}$, and suppose that $\neg Qualified([\texttt{ignite}])$ is true. The latter, in conjunction with axioms (3.7) and (3.8), implies that for all s, s' we have

$$\neg Result([\,], s) \vee \neg Holds(\neg disq(\texttt{ignite}), s) \vee \neg Successor(s, \texttt{ignite}, s') \tag{3.9}$$

The domain-independent axioms entail $\exists s\, [\, Result([\,], s) \wedge State(s)\,]$. Correctness of $\mathcal{FC}_\mathcal{D}$ moreover ensures that $State(s) \supset \forall s'.\, \neg Successor(s, \texttt{ignite}, s')$ is entailed iff so is $Holds(\texttt{runs}, s)$, for $\neg\texttt{runs}$ is the one and only regular precondition of action $\texttt{ignite}$. Put together, the disjunction (3.9) implies the following.

$$\forall s\, [\, Result([\,], s) \supset Holds(disq(\texttt{ignite}), s) \vee Holds(\texttt{runs}, s)\,]$$

That is to say, $\texttt{ignite}$ being unexecutable, i.e., $\neg Qualified([\texttt{ignite}])$, implies either an abnormal disqualification or that the engine is already running. ■

On the analogy of a result in Chapter 2, a one-to-one correspondence holds between the classical models of $\mathcal{FC}_\mathcal{D} \cup \{(3.7),(3.8),(2.37)\text{–}(2.40)\}$ and the interpretations for qualification scenarios of domain $\mathcal{D}$. More precisely, let ι be some model of the aforementioned collection of classical formulas and $I = (\Sigma, Res)$ an interpretation for a scenario $(\mathcal{O}, \mathcal{D})$. Then we say that ι and I *correspond* iff for all action sequences a^*, states S, and collections of fluent literals s such that $EUNA \models s = \tau_S$, we find that

$$Res(a^*) = S \;\text{ iff }\; [Result(a^*, s)]^\iota \text{ is true}$$

On this basis we can prove the following:

Theorem 3.6.1. *Let $(\mathcal{O}, \mathcal{D})$ be a qualification scenario with transition model Σ and Fluent Calculus axiomatization $\mathcal{FC}_\mathcal{D}$. Then for each model ι of $\mathcal{FC}_\mathcal{D} \cup \{(3.7),(3.8),(2.37)\text{–}(2.40)\}$ there exists a corresponding interpretation (Res, Σ) for $(\mathcal{O}, \mathcal{D})$ and vice versa.*

Proof. We confine ourselves to the differences to the proof of Theorem 2.9.2, which arise from the refined notion of qualified action sequences. The modified axiom (3.8) additionally requires that the fluent $disq(a)$ be false in the state resulting from performing a^*. This is exactly what distinguishes Definition 3.3.1 from Definition 2.7.2, regarding the notion of transition models. *Qed.*

From now onwards, let—given a qualification scenario $(\mathcal{O}, \mathcal{D})$ — $W_{(\mathcal{O},\mathcal{D})}$ denote the classical formulas $\mathcal{FC}_\mathcal{D} \cup \{(3.7),(3.8),(2.37)\text{–}(2.40)\}$ plus the axiomatization of the observations in $\mathcal{O}$.[9] Observations F after $[a_1, \ldots, a_n]$ are formalized as before, that is,

$$\exists s \, [\, Result([a_1, \ldots, a_n], s) \wedge Holds(F, s) \,] \tag{3.10}$$

while a disqualification observation a inexecutable after $[a_1, \ldots, a_n]$ is represented by

$$Qualified([a_1, \ldots, a_n]) \wedge \neg Qualified([a_1, \ldots, a_n, a]) \tag{3.11}$$

As one would expect, there is a one-to-one correspondence between the set of classical models of $W_{(\mathcal{O},\mathcal{D})}$ and the set of models of the scenario $(\mathcal{O}, \mathcal{D})$.

Corollary 3.6.1. *Let $(\mathcal{O}, \mathcal{D})$ be a qualification scenario with axiomatization $W_{(\mathcal{O},\mathcal{D})}$, then for each model ι of $W_{(\mathcal{O},\mathcal{D})}$ there exists a corresponding model (Res, Σ) of $(\mathcal{O}, \mathcal{D})$ and vice versa.*

Proof. Following Theorem 3.6.1 it suffices to show that an observation is true in Res iff ι is model of the observation's axiomatization.

1. By definition, an observation F after $[a_1, \ldots, a_n]$ is true in Res iff $Res([a_1, \ldots, a_n])$ is defined and formula F holds in that state. This in turn is equivalent to ι being a model of formula (3.10) according to axiom (2.38), which stipulates that $\exists s.\, Result(a^*, s)$ iff $Qualified(a^*)$; Proposition 2.9.3, which asserts that $Holds$ is suitably defined; and the fact that ι and (Σ, Res) correspond.
2. By definition, an observation a inexecutable after $[a_1, \ldots, a_n]$ is true in Res iff $Res([a_1, \ldots, a_n])$ is defined but $Res([a_1, \ldots, a_n, a])$ is not. This in turn is equivalent to ι being a model of formula (3.11) according to the fact that ι and (Σ, Res) correspond.

Qed.

[9] The reason for the at first glance unmotivated use of the letter W reveals in the following section.

3.6.2 Introducing Nonmonotonicity

The entailment relation of an action theory being a nonmonotonic one has been shown a fundamental characteristics of the Qualification Problem. Any axiomatization suitable for qualification domains and scenarios must therefore be based on some nonmonotonic extension to classical logic. Speaking less abstractly, this going beyond classical logic is necessary in order that abnormalities can be minimized, i.e., assumed away by default. We will meet this requirement by embedding our current Fluent Calculus-based axiomatization into a so-called default theory. Hence the general nonmonotonic framework to be employed is *Default Logic*, or rather, to be more precise, a conceptual extension of the original approach called *Prioritized Default Logic*. The latter is vital for reflecting possible degrees of abnormality when minimizing. For a formal introduction to both Default Logic and its prioritized variant see Annotation 3.3.

Given a qualification scenario $(\mathcal{O}, \mathcal{D})$, on the basis of the Fluent Calculus-axiomatization $W_{(\mathcal{O},\mathcal{D})}$ we construct a prioritized default theory as follows. The classical logic formulas $W_{(\mathcal{O},\mathcal{D})}$ constitute the background knowledge (which, by the way, resolves the mystery of why this denotation has been chosen). The various assumptions of 'normality' are formalized as default rules. For each abnormality fluent $f_{ab} \in \mathcal{F}_{ab}$, let $\delta_{f_{ab}}$ denote the default rule

$$\frac{: \forall s\, [\, Result([\,], s) \supset \neg Holds(f_{ab}, s)\,]}{\forall s\, [\, Result([\,], s) \supset \neg Holds(f_{ab}, s)\,]} \tag{3.12}$$

which is used to express the default assumptions that f_{ab} be false initially. That is to say, as long as it is consistent to assume that f_{ab} does not hold in the initial state, we do make this assumption. Let $D_{\mathcal{D}}$ denote the set of all default rules thus obtained from domain $\mathcal{D}$. Furthermore, let a partial ordering $<_{\mathcal{D}}$ be defined on $D_{\mathcal{D}}$ so that $\delta_{f_{ab}} <_{\mathcal{D}} \delta_{f_{ab}'}$ whenever $f_{ab} < f_{ab}'$ according to the partial ordering defined on the set $\mathcal{F}_{ab}$ of abnormality fluents.

This completes the axiomatization of qualification scenarios $(\mathcal{O}, \mathcal{D})$ as prioritized default theories $\Delta_{(\mathcal{O},\mathcal{D})} = (W_{(\mathcal{O},\mathcal{D})}, D_{\mathcal{D}}, <_{\mathcal{D}})$. Before we enter the proof of correctness, let us check on it by an example.

Example 3.6.2. Let $\mathcal{D}$ be the qualification domain of Example 3.2.1. Then $D_{\mathcal{D}}$ consists of the default rules $\delta_{\texttt{in(pot)}}$, $\delta_{\texttt{mysterious(ignite)}}$, and $\delta_{disq(\texttt{ignite})}$:

$$\begin{gathered}
\frac{: \forall s\, [\, Result([\,], s) \supset \neg Holds(\texttt{in(pt)}, s)\,]}{\forall s\, [\, Result([\,], s) \supset \neg Holds(\texttt{in(pt)}, s)\,]} \\
\frac{: \forall s\, [\, Result([\,], s) \supset \neg Holds(\texttt{mysterious(ignite)}, s)\,]}{\forall s\, [\, Result([\,], s) \supset \neg Holds(\texttt{mysterious(ignite)}, s)\,]} \\
\frac{: \forall s\, [\, Result([\,], s) \supset \neg Holds(disq(\texttt{ignite}), s)\,]}{\forall s\, [\, Result([\,], s) \supset \neg Holds(disq(\texttt{ignite}), s)\,]}
\end{gathered} \tag{3.13}$$

where $\delta_{\texttt{in(pot)}} <_{\mathcal{D}} \delta_{\texttt{mysterious(ignite)}}$ due to $\texttt{in(pt)} < \texttt{mysterious(ignite)}$.

The purpose of Default Logic is to extend classical logic, which is taken to encode precise knowledge, by expressions that formalize somehow vague, defeasible knowledge. Called *default rules* (or *defaults*, for short), these expressions allow for stating that some property α 'normally' implies some property ω. The reference to normality is made precise by specifying circumstances $\neg\beta$ which must provably hold in order that the conclusion from α to ω shall not be valid. The formal syntax of a default is $\frac{\alpha:\beta}{\omega}$ where α (the *prerequisite*), β (the *justification*), and ω (the *consequence*) all are formulas in classical logic. For our purpose, it suffices to consider only so-called normal defaults, where justification and consequence coincide. In case one or more components of a default contain free variables, the then so-called open default is considered representative for all of its ground instances. A *default theory* $\Delta = (W, D)$ consists of a set of defaults D and a set of closed formulas W, the latter of which is called *world-* (or *background*) *knowledge*. Reasoning in default theories is based on the formation of so-called extensions. These are obtained by starting with the background knowledge, W, and successively applying defaults chosen from D, which amounts to adding their consequences provided their justification is consistent with what is finally obtained as extension. Once there are no more applicable defaults left, the deductive closure of the resulting set of formulas constitutes an extension. An extension may be regarded as one coherent view on the state of affairs. Formally, let E be a set of closed formulas. We define

1. $\Gamma_0 = W$
2. $\Gamma_i = Th(\Gamma_{i-1}) \cup \{\omega : \frac{\alpha:\omega}{\omega} \in D,\ \alpha \in \Gamma_{i-1},\ \neg\omega \notin E\}$, for $i > 0$.

(where $Th(\Psi)$ denotes the (classical) deductive closure of Ψ). Then E is an *extension* of Δ iff $E = \bigcup_{i=0}^{\infty} \Gamma_i$. A default theory may admit multiple extensions, each of which is obtained by applying different subsets of the underlying defaults. Then a closed formula is said to be *skeptically entailed* in a default theory iff it is contained in all extensions of the latter.

When constructing extensions of a default theory in the way just described, all defaults are applied with the same priority. For applications where this is undesired, an extension of classical Default Logic called *Prioritized Default Logic* supports the specification of (possibly partial) preference orderings among defaults. This ordering is exploited to select among the extensions of a default theory those in which the most preferred defaults have been applied. Formally, a *prioritized default theory* is a triple $\Delta = (W, D, <)$ where W and D are as in standard Default Logic and $<$ is a partial ordering on D. If E is a closed set of formulas, then a default $\frac{\alpha:\omega}{\omega}$ is said to be *applied in* E iff $\alpha, \omega \in E$. An extension E of the (standard) default theory (W, D) is a *prioritized extension* of Δ iff there is a strict ordering $\ll$ extending $<$ such that the following holds for all extensions E' of (W, D) and all defaults $\delta' \in D$: If δ' is applied in $E' \setminus E$, then there is some $\delta \ll \delta'$ which is applied in $E \setminus E'$. In words, a standard extension E is prioritized if we can find a strict ordering respecting $<$ and the following is true: Whenever some default δ' is not applied in E but in some other standard extension E', then there is also a default δ which is applied in E but not in E' and which has higher priority than δ' according to the strict ordering.

Annotation 3.3. Classical and Prioritized Default Logic.

Now, suppose $\mathcal{O}$ consists of the observations

$$\neg\mathtt{runs}\ \underline{\text{after}}\ [\,]$$

$$\mathtt{ignite}\ \underline{\text{inexecutable}}\ \underline{\text{after}}\ [\,]$$

We have already seen that $W_{(\mathcal{O},\mathcal{D})}$ and the axiomatization of the second observation, viz. $\mathit{Qualified}([\,]) \wedge \neg \mathit{Qualified}([\mathtt{ignite}])$, entail

$$\forall s\,[\,\mathit{Result}([\,],s) \supset \mathit{Holds}(\mathit{disq}(\mathtt{ignite}),s) \vee \mathit{Holds}(\mathtt{runs},s)\,]$$

The encoding of the first observation, viz. $\exists s\,[\,\mathit{Result}([\,],s) \wedge \mathit{Holds}(\neg\mathtt{runs},s)\,]$, in conjunction with the axiom $\mathit{Result}([\,],s) \wedge \mathit{Result}([\,],s') \supset s = s'$ allows to strengthen the above to

$$\forall s\,[\,\mathit{Result}([\,],s) \supset \mathit{Holds}(\mathit{disq}(\mathtt{ignite}),s)\,]$$

It follows that default $\delta_{\mathit{disq}(\mathtt{ignite})}$ cannot be applied. From the state constraint $\mathit{disq}(\mathtt{ignite}) \equiv \exists x.\,\mathtt{in}(x) \vee \mathtt{mysterious(ignite)}$ in $\mathcal{FC}_\mathcal{D}$ we also conclude that

$$\forall s\,[\,\mathit{Result}([\,],s) \supset \exists x.\,\mathit{Holds}(\mathtt{in}(x),s) \vee \mathit{Holds}(\mathtt{mysterious(ignite)},s)\,]$$

Since entity $\mathtt{pt}$ is the only one, it follows that $\delta_{\mathtt{in(pot)}}$ and $\delta_{\mathtt{mysterious(ignite)}}$ are mutually exclusive, for either consequence in conjunction with $W_{(\mathcal{O},\mathcal{D})}$ implies the negation of the other rule's justification. According to the priority ordering $<_\mathcal{D}$, default $\delta_{\mathtt{in(pot)}}$ is to be preferably applied when seeking prioritized extensions of $\Delta_{(\mathcal{O},\mathcal{D})}$. Our default theory $\Delta_{(\mathcal{O},\mathcal{D})}$ thus admits a unique prioritized extension E, which includes the following formula.

$$\forall s[\mathit{Result}([\,],s) \supset s = \neg\mathtt{runs} \circ \mathtt{in(pt)} \circ \neg\mathtt{mysterious(ignite)} \circ \mathit{disq}(\mathtt{ignite})]$$

This formula is skeptically entailed by $\Delta_{(\mathcal{O},\mathcal{D})}$. The reader may notice that accordingly the qualification scenario $(\mathcal{O},\mathcal{D})$ admits a unique preferred model (Σ,Res), which satisfies

$$\mathit{Res}([\,]) = \{\neg\mathtt{runs}, \mathtt{in(pt)}, \neg\mathtt{mysterious(ignite)}, \mathit{disq}(\mathtt{ignite})\}$$

■

In the remainder of this section, we prove general correctness of our axiomatization of qualification scenarios by means of Prioritized Default Logic. Some preparations are required to this end. Let $\Delta_{(\mathcal{O},\mathcal{D})}$ be the axiomatization of a qualification scenario $(\mathcal{O},\mathcal{D})$ with $\mathcal{F}_{ab}$ being the abnormality fluents. Then a prioritized extension E of $\Delta_{(\mathcal{O},\mathcal{D})}$ and an interpretation (Σ,Res) for this scenario are said to *correspond* iff for all $f_{ab} \in \mathcal{F}_{ab}$ we find that

$$f_{ab} \in \mathit{Res}([\,]) \text{ iff } \forall s\,[\,\mathit{Result}([\,],s) \supset \neg\mathit{Holds}(f_{ab},s)\,] \in E \qquad (3.14)$$

The notion of *potential extensions* to be introduced next will be crucial for the proof of our main theorem. In what follows, for notational convenience we use the abbreviation $\mathit{Initially}(\neg\ell)$ for $\forall s\,[\,\mathit{Result}([\,],s) \supset \neg\mathit{Holds}(\ell,s)\,]$. Suppose F is a set of formulas which consists in the following:

1. $W_{(\mathcal{O},\mathcal{D})}$;
2. either $\mathit{Initially}(\neg f_{ab})$ or $\neg\mathit{Initially}(\neg f_{ab})$, for each abnormality fluent f_{ab}.

Then $E = \mathit{Th}(F)$ is called a potential extension of $\Delta_{(\mathcal{O},\mathcal{D})}$. Notice that potential extensions may be inconsistent, e.g., if $\mathit{Initially}(\neg f_{ab}) \in F$ although $W_{(\mathcal{O},\mathcal{D})}$ necessitates that f_{ab} be true in any s which satisfies $\mathit{Result}([\,], s)$.

Given a potential extension $E = \mathit{Th}(F)$, we call *induced by* E any strict ordering $\ll_{\mathcal{D}}$ that extends $<_{\mathcal{D}}$ such that

$$\delta_{f_{ab}} \ll_{\mathcal{D}} \delta_{f_{ab}'} \text{ whenever } \mathit{Initially}(\neg f_{ab}) \in F \text{ and } \neg\mathit{Initially}(\neg f'_{ab}) \in F$$

Induced orderings will be used below to verify the constituent properties for potential extensions being prioritized extensions. It is easy to verify that the standard extensions of the default theory $(W_{(\mathcal{O},\mathcal{D})}, D_{\mathcal{D}})$ are always potential extensions.

Lemma 3.6.1. *Let* $\Delta_{(\mathcal{O},\mathcal{D})} = (W_{(\mathcal{O},\mathcal{D})}, D_{\mathcal{D}}, <_{\mathcal{D}})$ *be the axiomatization of some qualification scenario, then each (standard) extension of* $(W_{(\mathcal{O},\mathcal{D})}, D_{\mathcal{D}})$ *is a potential extension of* $\Delta_{(\mathcal{O},\mathcal{D})}$.

Proof. Let E be an extension of $(D_{\mathcal{D}}, W_{(\mathcal{O},\mathcal{D})})$, and let

1. $\Gamma_0 = W_{(\mathcal{O},\mathcal{D})}$;
2. $\Gamma_1 = \mathit{Th}(\Gamma_0) \cup \{\omega : \frac{:\omega}{\omega} \in D_{\mathcal{D}},\ \neg\omega \notin E\}$; and
3. $\Gamma_2 = \mathit{Th}(\Gamma_1)$.

Since all possibly applicable defaults in $D_{\mathcal{D}}$ have been applied to compute Γ_1 and since E is extension, we know that $\Gamma_2 = E$. By construction, Γ_2, hence E, is subset of some potential extension. To see, then, why it equals a potential extension, observe first that $E = \mathit{Th}(E)$ and $W_{(\mathcal{O},\mathcal{D})} \in E$. It remains to verify that for every $f_{ab} \in \mathcal{F}_{ab}$, E includes either $\mathit{Initially}(\neg f_{ab})$ or $\mathit{Initially}(\neg f_{ab})$. Let $f_{ab} \in \mathcal{F}_{ab}$ be an abnormality fluent. From $\frac{:\mathit{Initially}(\neg f_{ab})}{\mathit{Initially}(\neg f_{ab})} \in D_{\mathcal{D}}$ and the construction of Γ_1, we know that either $\mathit{Initially}(\neg f_{ab}) \in \Gamma_1$, hence $\mathit{Initially}(\neg f_{ab}) \in E$, or else $\neg\mathit{Initially}(\neg f_{ab}) \in E$. *Qed.*

The notion of correspondence between interpretations for qualification scenarios and prioritized extensions generalizes to potential extension in the obvious way—a potential extension $E = \mathit{Th}(F)$ and an interpretation (Σ, Res) correspond iff the condition (3.14) holds for all $f_{ab} \in \mathcal{F}_{ab}$. Notice that each interpretation has a unique corresponding potential extension, whereas there may be multiple interpretations corresponding to a single potential extension. For the latter does not necessarily fix all states resulting from the performance of action sequences. Notice further that whenever E is consistent then there exists a corresponding interpretation which is a model of the qualification scenario at hand. This is granted by Corollary 3.6.1, for if E is consistent it admits a (classical) model ι—which then corresponds to some model of the scenario.

We are now prepared to prove general correctness of our entire axiomatization.

Theorem 3.6.2. *Let* $(\mathcal{O}, \mathcal{D})$ *be a qualification scenario whose axiomatization is the prioritized default theory* $\Delta_{(\mathcal{O},\mathcal{D})} = (W_{(\mathcal{O},\mathcal{D})}, D_{\mathcal{D}}, <_{\mathcal{D}})$*, then for each prioritized extension of* $\Delta_{(\mathcal{O},\mathcal{D})}$ *there exists a corresponding preferred model of* $(\mathcal{O}, \mathcal{D})$ *and vice versa.*

Proof. Let $\mathcal{F}_{ab}$ be the set of abnormality fluents of $\mathcal{D}$.

"$\Leftarrow$":

Let $M = (\Sigma, Res)$ be some preferred model of $(\mathcal{O}, \mathcal{D})$, and let $E = Th(F)$ be the potential extension corresponding to M. To begin with, we prove that E is a standard extension of $(W_{(\mathcal{O},\mathcal{D})}, D_{\mathcal{D}})$. Let

1. $\Gamma_0 = W_{(\mathcal{O},\mathcal{D})}$;
2. $\Gamma_1 = Th(\Gamma_0) \cup \{\omega : \frac{:\omega}{\omega} \in D_{\mathcal{D}},\ \neg\omega \notin E\}$; and
3. $\Gamma_2 = Th(\Gamma_1)$.

Then we have to verify that $\Gamma_2 = E$ (c.f. the proof of Lemma 3.6.1). Clearly, $\Gamma_2 \subseteq E$, since for any $\frac{:\omega}{\omega} \in D_{\mathcal{D}}$ such that $\omega \in \Gamma_1$, we have $\neg\omega \notin E$, which in turn implies $\omega \in E$ given that E is a potential extension. Moreover, the assumption $\Gamma_2 \subsetneq E$ leads to a contradiction: Suppose $\Gamma_2 \subsetneq E$, then this indicates the existence of some $\frac{:\omega}{\omega} \in D_{\mathcal{D}}$ (where $\omega = \mathit{Initially}(\neg f_{ab})$ for some $f_{ab} \in \mathcal{F}_{ab}$) such that $\neg\omega \in E$ but $\neg\omega \notin \Gamma_2$. Let Ω be the set of all these ω, i.e., $\Omega = \{\neg\omega \in E : \neg\omega \notin \Gamma_2\}$. Then $E' = (E \setminus \Omega) \cup \{\omega : \neg\omega \in \Omega\}$ is an extension of $(W_{(\mathcal{O},\mathcal{D})}, D_{\mathcal{D}})$. From Lemma 3.6.1 we conclude that E' is a potential extension of $\Delta_{(\mathcal{O},\mathcal{D})}$. Let M' be an interpretation corresponding to E' such that M' is a model of $(\mathcal{O}, \mathcal{D})$. From the construction of E' and the definition of correspondence it follows that M' contains strictly less abnormality assumptions than M, given that Ω is non-empty. Hence, $M' \prec M$, which contradicts M being a preferred model.

Having proved that E is an extension of $(W_{(\mathcal{O},\mathcal{D})}, D_{\mathcal{D}})$, it remains to be shown that it is prioritized. Let $\ll_{\mathcal{D}}$ be an arbitrary strict ordering induced by E. Furthermore, let E' be any extension of $(W_{(\mathcal{O},\mathcal{D})}, D_{\mathcal{D}})$ and M' be an interpretation corresponding to E' and which is a model of $(\mathcal{O}, \mathcal{D})$. Suppose $\delta_{f_{ab}'} \in D_{\mathcal{D}}$ is a default which is applied in $E' \setminus E$. Then $\neg f_{ab}' \in Res'([\,])$ but $f_{ab}' \in Res([\,])$. Model M being preferred, we know that $M' \not\prec M$. Therefore, we can also find some $f_{ab} \in \mathcal{F}_{ab}$ such that $\neg f_{ab} \in Res([\,])$ but $f_{ab} \in Res'([\,])$. It follows that $\delta_{f_{ab}} \ll_{\mathcal{D}} \delta_{f_{ab}'}$ (since $\ll_{\mathcal{D}}$ is induced by E), that is, there exists a default which is preferred (wrt. $\ll_{\mathcal{D}}$) to $\delta_{f_{ab}'}$ and which is applied in $E \setminus E'$.

"$\Rightarrow$":

Let E be a prioritized extension of $\Delta_{(\mathcal{O},\mathcal{D})}$. Then E is an extension of $(W_{(\mathcal{O},\mathcal{D})}, D_{\mathcal{D}})$ and also, according to Lemma 3.6.1, a potential extension. Let M be an interpretation corresponding to E such that M is a model of $(\mathcal{O}, \mathcal{D})$. By contradiction, we prove that M is preferred. Suppose there exists a preferred model M' of $(\mathcal{O}, \mathcal{D})$ such that $M' \prec M$. This implies the existence of a corresponding prioritized extension E' of $\Delta_{(\mathcal{O},\mathcal{D})}$ according to the first part ("$\Leftarrow$") of this proof. From $M' \prec M$ we conclude that there exists some $\delta_{f_{ab}}{'} \in D_{\mathcal{D}}$ which is applied in $E' \setminus E$ but no $\delta_{f_{ab}} \in D_{\mathcal{D}}$ with higher priority and which is applied in $E \setminus E'$. This contradicts E being prioritized extension. *Qed.*

An immediate consequence of this one-to-one correspondence is that, as far as observations are concerned, the notion of skeptical entailment in prioritized default theories resulting from our axiomatization and the notion of entailment suggested by our action theory coincide.

Corollary 3.6.2. *Let $(\mathcal{O}, \mathcal{D})$ be a qualification scenario with axiomatization $\Delta_{(\mathcal{O},\mathcal{D})}$. An observation is entailed by $(\mathcal{O}, \mathcal{D})$ iff the corresponding formula (c.f. (3.10) and (3.11), respectively) is skeptically entailed by $\Delta_{(\mathcal{O},\mathcal{D})}$.*

Proof. The claim is a consequence of Theorem 3.6.2 following the lines of the proof of Corollary 3.6.1. *Qed.*

3.7 Bibliographic Remarks

The Qualification Problem was introduced and so named in [76] as one of several arguments making manifest the urge for nonmonotonic representation and reasoning frameworks. The very article already anticipated the solution of introducing abnormality predicates which are to be (globally) assumed away by default. This solution was formally elaborated in [78] based on the by then existing nonmonotonic formalism of so-called circumscription [77]. It has first been observed in [64], however, that globally minimizing abnormal disqualifications of actions fails to suitably account for disqualifications that occur for reasons of causality.[10] Rather than offering a solution, however,

[10] It is remarkable that the proposal put forth in the very article [78] to address the Frame Problem by globally minimizing change has been proved erroneous by a counter-example that shows some striking similarities to the refutation of global minimization as means to tackle the Qualification Problem. Suppose we consider abnormal any change of a fluent's truth value during the execution of an action, as suggested in [78]. Then the *Yale Shooting* problem, which we already touched upon earlier in this book, arises as follows (c.f. [49]): Given that shooting at a turkey with a loaded gun causes the former to drop dead, we should expect exactly this to happen when we load the gun, wait for a short period, and then shoot. Yet globally minimizing abnormalities in this scenario produces a second model where the gun surprisingly becomes unloaded during the intermediary action of waiting and the turkey survives! While the magical change of the gun's status is abnormal, the turkey surviving the shot is normal in the above sense—as opposed to the change of its life status in the intended model. Hence, this second model minimizes abnormality as well, though it is obviously counter-intuitive. The problem here is essentially that uniformly considering abnormal all changes is ill-defined. A gun that becomes magically unloaded while waiting deserves being called abnormal but not the death of the turkey if being shot at with a loaded gun, which is perfectly normal from

this paper marked the beginning of a shift away from the original concept of the Qualification Problem. Instead of being concerned with unlikely action preconditions to be assumed away by default, the Qualification Problem there has been taken as the task of extracting implicit action preconditions from general knowledge, e.g., state constraints—a task which we have already considered in the context of the Ramification Problem (see Section 2.8). The fundamental difference to the plain Qualification Problem is that any such implicit condition needs to be explicitly verified, hence cannot be assumed away prior to concluding that the action in question is executable. This reinterpretation of the Qualification Problem has been taken up in subsequent work, e.g., [38, 68, 99]. There is no urge for arguing against the view that accounting for implicit action preconditions is part of some broader Qualification Problem. But denying that the core is to providing means to assume away, by default, the occurrence of *a priori* unlikely obstacles amounts to an oversimplification of the problem.

One of the marginal consequences of dealing with the Qualification Problem is that a property called "restricted monotonicity," which has been claimed generally desirable for theories of actions in [67], is no longer so when facing qualification scenarios. A formalism possesses this property if additional observations can only increase the set of observations that are entailed by a domain description. This kind of monotonicity is obviously not appropriate in case of observations following by default.

Exploiting solutions to the Ramification Problem when addressing the problem of caused vs. unmotivated action disqualifications has first been proposed in [110]. An earlier general alternative to global minimization, namely, *chronological ignorance* [102, 103], is in principle capable of treating correctly our key example with the potato being deliberately placed in the tail pipe (viz. Example 3.2.2), but the approach suffers from another, inherent limitation. Roughly speaking, the crucial idea underlying the principle of chronological ignorance is to assume away, by default, abnormal circumstances but to prefer minimization of abnormalities at earlier timepoints.[11] Formally, a certain kind of modal logic is employed as a means to express the distinction between provable facts and propositions which might or might not be true. Our example domain, for instance, one would formulate in this framework by these two action descriptions:

$$\Box\,\textsc{True}\,(\texttt{insert}(x),t) \land \Box\,\textsc{True}\,(\neg\texttt{in}(x),t) \land \Diamond\,\textsc{True}\,(\neg\texttt{heavy}(x),t) \supset \Box\,\textsc{True}\,(\texttt{in}(x),t+1) \tag{3.15}$$

the perspective of causality. The strong resemblance is apparent to the observation that a potato being too heavy to lift is truly abnormal as opposed to the expectation that one will fail to start the car after attempting to place a potato in the tail pipe.

[11] This explains the naming: Potential abnormal disqualifications are *ignored* whenever possible, and this is done in *chronological* order. Assuming away obstacles whenever their occurrence cannot be proved, Shoham also calls the "ostrich" principle, or: "what-you-don't-know-won't-hurt-you."

$$\Box\,\mathrm{TRUE}\,(\mathtt{ignite}, t) \wedge \Box\,\mathrm{TRUE}\,(\neg\mathtt{runs}, t) \wedge \Diamond\forall x.\ \mathrm{TRUE}\,(\neg\mathtt{in}(x), t) \\ \supset \Box\,\mathrm{TRUE}\,(\mathtt{runs}, t+1) \tag{3.16}$$

where $\Box\,\mathrm{TRUE}\,(\ell, t)$ should be read as "at time t fluent literal ℓ provably holds" and $\Diamond\,\mathrm{TRUE}\,(\ell, t)$ as "at time t fluent literal ℓ may or may not hold." Thus the first of the two implications states that if it is known that some action $\mathtt{insert}(x)$ occurs at time t, that $\neg\mathtt{in}(x)$ holds, and that possibly $\neg\mathtt{heavy}(x)$ holds at that time, then $\mathtt{in}(x)$ holds at time $t+1$. Likewise, if it is known that the action $\mathtt{ignite}$ occurs at time t, that $\neg\mathtt{runs}$ holds, and that possibly $\forall x.\neg\mathtt{in}(x)$ holds at that time, then $\mathtt{runs}$ holds at time $t+1$. Observe how regular action preconditions, like $\neg\mathtt{in}(x)$ and $\neg\mathtt{runs}$, must provably hold whereas abnormal disqualifying conditions, like $\mathtt{heavy}(x)$ and $\exists x.\,\mathtt{in}(x)$, are assumed away whenever the opposite does not provably hold. Now, suppose given $\Box\,\mathrm{TRUE}\,(\neg\mathtt{in}(\mathtt{pt}), 1) \wedge \Box\,\mathrm{TRUE}\,(\neg\mathtt{runs}, 1)$ in conjunction with the action occurrences $\Box\,\mathrm{TRUE}\,(\mathtt{insert}(\mathtt{pt}), 1) \wedge \Box\,\mathrm{TRUE}\,(\mathtt{ignite}, 2)$. Then chronological ignorance tells us that $\Diamond\,\mathrm{TRUE}\,(\neg\mathtt{heavy}(\mathtt{pt}), 1)$ holds since nothing is known about $\mathrm{TRUE}\,(\neg\mathtt{heavy}(\mathtt{pt}), 1)$ itself. Hence, (3.15) implies $\Box\,\mathrm{TRUE}\,(\mathtt{in}(\mathtt{pt}), 2)$, which gives us $\neg\Diamond\forall x.\ \mathrm{TRUE}\,(\neg\mathtt{in}(x), 2)$.[12] Thus the antecedent of the implication (3.16) is false (for $t = 2$) and, consequently, the second action, $\mathtt{ignite}$, is correctly concluded unqualified. Notice that this being the unique suggested course of events relies on the chronological order in which minimization is performed. Otherwise, it could equally well be concluded that $\Diamond\forall x.\ \mathrm{TRUE}\,(\neg\mathtt{in}(x), 2)$ holds, for, in the first place, nothing is known about $\forall x.\ \mathrm{TRUE}\,(\neg\mathtt{in}(x), 2)$ itself. This in turn entails $\neg\Diamond\,\mathrm{TRUE}\,(\neg\mathtt{heavy}(\mathtt{pt}), 1)$, i.e., $\Box\,\mathrm{TRUE}\,(\mathtt{heavy}(\mathtt{pt}), 1)$, since (3.15) is logically equivalent to

$$\Box\,\mathrm{TRUE}\,(\mathtt{insert}(x), t) \wedge \Box\,\mathrm{TRUE}\,(\neg\mathtt{in}(x), t) \wedge \Diamond\,\mathrm{TRUE}\,(\neg\mathtt{in}(x), t+1) \\ \supset \Box\,\mathrm{TRUE}\,(\mathtt{heavy}(x), t)$$

This alternative conclusion corresponds to the counter-intuitive model obtained by global minimization of abnormalities (c.f. Section 3.2) but, as indicated, it is not supported by chronological ignorance.

The interesting, albeit informal, reason for chronological ignorance coming to the desired conclusion in this and similar cases is a certain respect of causality hidden in this method. By minimizing chronologically, one tends to minimize causes rather than effects, which is the right thing to do, simply because causes generally precede their effects. On the other hand, the applicability of chronological minimization is known to be intrinsically restricted to domains and scenarios which do not involve indeterminate information. This has been shown for a variety of aspects of non-determinism; see, e.g., [60, 88, 104]. Informally speaking, the problem reveals whenever indeterminate information provides sufficient evidence for an abnormality without, by virtue of not being deterministic, necessitating it. Putting off abnormalities for as long as possible then ignores uncertain evidence and, in so doing,

12 As usual in modal logic, $\neg\Box\,\mathrm{TRUE}\,(\ell, t)$ is equivalent to $\Diamond\,\mathrm{TRUE}\,(\neg\ell, t)$.

supports unsound conclusions. For illustration purpose, let us formalize a scenario mentioned by [88].

Example 3.7.1. Suppose we park our car overnight in a neighborhood that is known for its suffering from a tail pipe marauder. Then chances are good that the rascal has struck by the next morning so that we had better check for a potato in the tail pipe before trying to start the car. Let $\mathcal{D}$ be the qualification domain of Example 3.2.2 but with action name `insert` replaced by $\texttt{stay-overnight}^0$. The latter having a non-deterministic effect, it is specified by a pair of non-exclusive action laws, viz.

$$\begin{array}{llll} \texttt{stay-overnight} & \underline{\text{transforms}}\ \{\} & \underline{\text{into}} & \{\} \\ \texttt{stay-overnight} & \underline{\text{transforms}}\ \{\neg\texttt{in(pt)}\} & \underline{\text{into}} & \{\texttt{in(pt)}\} \end{array}$$

Let $\mathcal{O}$ consist of the observation $\neg\texttt{runs} \wedge \neg\texttt{in(pt)}\ \underline{\text{after}}\ [\,]$, then the qualification scenario $(\mathcal{O}, \mathcal{D})$ admits two categories of preferred models (Σ, Res), one of which satisfy $\neg\texttt{in(pt)} \in \mathit{Res}([\texttt{stay-overnight}])$ while the others claim that $\texttt{in(pt)}$ be true in $\mathit{Res}([\texttt{stay-overnight}])$. Hence nothing is entailed as to whether there is an abnormal disqualification of `ignite` or not after staying overnight. ∎

Now to the problems of chronologically minimizing in this scenario. The following formula specifies the new, non-deterministic action:

$$\begin{array}{l} \Box\,\textsc{True}\,(\texttt{stay-overnight}, t) \wedge \Box\,\textsc{True}\,(\neg\texttt{in(pt)}, t) \\ \quad \supset \Box\,\textsc{True}\,(\texttt{in(pt)}, t+1) \vee \Box\,\textsc{True}\,(\neg\texttt{in(pt)}, t+1) \end{array} \tag{3.17}$$

That is to say, if at time t we stay overnight and there is no potato in the tail pipe, then at time $t+1$ this may or may not have changed. Consider this formula in conjunction with the specification of action `ignite` as given by implication (3.16), and suppose given $\Box\,\textsc{True}\,(\neg\texttt{in(pt)}, 1)$ in conjunction with the events $\Box\,\textsc{True}\,(\texttt{stay-overnight}, 1)$ and $\Box\,\textsc{True}\,(\texttt{ignite}, 2)$. Then implication (3.17) allows no definite conclusion about $\textsc{True}\,(\texttt{in(pt)}, 2)$, hence $\Diamond\,\textsc{True}\,(\neg\texttt{in(pt)}, 2)$ following the principle of chronological ignorance, which in turn implies $\Box\,\textsc{True}\,(\texttt{runs}, 3)$ according to (3.16). Thus the conclusion that `ignite` be qualified at time 2 is sanctioned despite the possibility that the tail pipe marauder has struck by then. This undesired conclusion proves that the "what-you-don't-know-won't-hurt-you" principle is not suited for indeterminate information. While the Qualification Problem requires assuming away abnormal circumstances whenever they do not provably hold, this is in general too credulous if the performance of a non-deterministic action renders quite possible such circumstances. A "what-you-can't-expect-won't hurt-you" principle would be appropriate—clearly, an abnormal disqualification of `ignite` after carelessly leaving unattended the car in the dangerous neighborhood is to be expected from one of the possible effects of staying overnight, which is why this potential disqualification ought not be assumed away.

Exploiting ramification as the key to the Qualification Problem shares with *Motivated Action Theory* (MAT, for short) [104, 2] the basic insight that an appropriate notion of causality is necessary when assuming away abnormalities. In the latter framework, event happenings are minimized while taking into account the possibility that their occurrence is being caused (or, in other words, *motivated*, hence the name). This excluding unmotivated events and our minimizing uncaused abnormal disqualifications are somehow complementary while being based on the same principles. However, some methodical criticism and limitations apply to MAT: An unsatisfactory property of the underlying preference criterion, i.e., motivation, is its depending on the syntactical structure of the formulas representing causal knowledge. As a consequence, logically equivalent formalizations may induce different preference criteria, of which only one is the desired. Moreover, the formal concept of motivation becomes rather complicated in case of disjunctive, i.e., indeterminate information, which entails difficulties with assessing the range of applicability of MAT in general. Finally, events can only be 'motivated' by past events, which does not allow for telling apart caused events that occur concurrently with the triggering event. (In passing, we mention that this last restriction also applies to a version of the Event Calculus (see below) if events are allowed to trigger other events, as proposed in [98]). In comparison, in [115] it is illustrated that ramification can also be successfully applied to the problem of minimizing event occurrences.

Talking about the *Event Calculus* (introduced by [61]; for an overview and thorough analysis see [100] and the monograph [99]), dealing with the Qualification Problem there requires to respect a fundamental difference between this approach and, say, both Situation Calculus and Fluent Calculus. Namely, the latter are based on a branching time structure where different, hypothetical action sequences may fork left and right of the actual time line, if any. In contrast, the Event Calculus grounds on a linear time structure thus representing solely the actual evolution of the world. The occurrence of actions is specified by assertions of the form $Happens(A, T)$ stating that action A is performed at time T. A consequence of the linear time structure is that once an assertion of this form is made, it cannot be withdrawn. This entails difficulties with formally representing observations of abnormal action disqualifications if these are interpreted as the action being physically impossible—which is the interpretation we pursued throughout our analysis of the Qualification Problem. An alternative view seems better suited for the Event Calculus or, for that matter, generally for action theories based on a linear time structure. Instead of tightly coupling an action with its success, such as starting the car with the effect that the engine is running, performing an action may first of all be taken as the mere intention to achieve a certain effect. The actual achievement then depends on both regular conditions being met and the absence of abnormal, disqualifying circumstances. We do not wish to provide a detailed formal account of the Qualification Problem in the Event Calculus here, but let us mention that this is accomplishable by introducing a predicate $Succeeds(a, t)$ meant to indicate that action a happening at time t produces the intended effect. Abnormal conditions denying the success of an action can then be dealt with along the lines of our action theory.

Default Logic [85], the formalism in which our Fluent Calculus axiomatization of Chapter 2 has been embedded in view of the Qualification Problem, was among the earliest nonmonotonic frameworks, the key reference to many of which is the special volume [11]. Numerous modifications and extensions to classical Default Logic have subsequently been developed in order to cope with a number of ontological aspects missing or arguably being counterintuitively dealt with in the original approach; a variant which uniformly addresses many of these aspects is presented in [19], just to mention one. However, the default rules occurring in our axiomatization bear a very specific structure so that, fortunately, none of the criticisms of classical Default Logic applies (speaking in technical terms, our defaults are, without exception, both "prerequisite-free" and "normal", which means they all are of the form $\frac{:\omega}{\omega}$). Prioritized Default Logic has been introduced in [15] to support the notion of preference among default rules. We have adopted the more elegant reformulation proposed by [87], who also points out some problems with the original approach for certain classes of default theories—but again our particular default theories are not subject to this criticism due to their special structure. The approach [13] provides the first formal framework for priorities among unrestricted default rules. A variety of approaches to the automation of reasoning in default logics exist, e.g. [85, 7, 97, 96], just to mention a few. Those calculi which perform proof search in a local fashion, that is, which do not necessarily require the generation of entire extensions of an underlying default theory, are usually, however, restricted to so-called credulous reasoning. In contrast to skeptical reasoning, the latter entails any formula that belongs to (at least) one extension. A general method for extending credulous proof procedures to skeptical entailment without loosing the property of being local and goal-oriented has been developed in the second part of [106].

For rounding off this chapter on the Qualification Problem some remarks seem appropriate on the topical debate about nonmonotonic vs. probabilistic logics. While the former amounts to qualitative reasoning about small likelihoods, the latter is concerned with quantitative reasoning on the basis of exact knowledge of probability values. The strongest argument undermining the fundamental of all nonmonotonic frameworks is that sometimes counterintuitive, if not contradictory, conclusions are unavoidable. This is best illustrated with the well-known *Lottery Paradox* introduced in [62] (see also [84]):

> "In a fair lottery with 100 tickets the chance that any given ticket will lose is 99 per cent. It therefore seems reasonable, for any given ticket, to believe, or *accept* as a basis for action, the statement 'This ticket will lose.' Yet the conjunction of such statements for all tickets must be false, since some ticket will win, so we can hardly accept the conjunction. We seem to have to accept each statement separately, but not the conjunction of them."[13]

[13] [63], pp. 187–8

Scenarios like the foregoing clearly favor the use of probabilistic theory when it comes to decision-making based on the available information. On the other hand, it seems that only for a small fraction of daily-life situations such exact (conditional) probabilities can be provided. This is the strongest argument against probabilistic reasoning as an alternative to nonmonotonic logics.[14] The relevance of this argument has revealed in the course of a research project devoted to formalizing large parts of human common sense knowledge:

> "Early on, we allowed each assertion in the knowledge base to have a numeric certainty factor, but this approach led to its own set of increasingly severe difficulties. [...] There wasn't statistical data to support these numbers [...]. These problems led us to go back to a simple nonnumeric scheme in which [assertions] [...] would simply be true (nonmonotonically, that is, true by default) [...]."[15]

It so seems that the pragmatic answer to the problem of which approach provides the 'right' formalism, nonmonotonic or probabilistic logic, is to compromise: Exact probability values should be used whenever both available and appropriate—as it is the case in the scenario described by the Lottery Paradox, for instance. When it comes to representing knowledge involving abnormalities which are only vaguely known to be unlikely, choosing a nonmonotonic framework helps to base decision-making on reasonable conclusions.[16]

---

[14] Another argument often brought forward, namely, that probabilistic reasoning is computationally intractable, has been challenged by the notable practical success of *Bayesian Belief Networks* [81], which exploit knowledge of causal independence to speed up the reasoning process.

[15] [46], pp. 34–5

[16] To complete this brief discussion, successful attempts to embed nonmonotonic reasoning in logics of probability should be mentioned, e.g., [42].

4. Qualified Ramifications

4.1 State Constraints with Exceptions

Our account of the Qualification Problem was made possible by the foregoing solution to the Ramification Problem. The fact that abnormal circumstances can be brought about as side effect of performing certain actions is best accommodated, so our argument went, if these side effects are obtained just like all indirect effects are, that is, through ramification. In this concluding chapter, we will expand even further the connection between the Ramification and Qualification Problem. This time it is argued how solutions to the latter, and in particular the one we pursued, can in turn be successfully employed to address a generalization of the former.

The motivation for this generalization of the Ramification Problem is inherited from the Qualification Problem. If actions in non-artificial environments may turn out unqualified for some abnormal reason, then so may ramifications, too. Recall, for instance, the introductory electric circuit from the beginning of Chapter 2, which connects a battery with a switch and a light bulb (c.f. Fig. 2.1). It has been said that the light is on if and only if the switch is closed. Formally, $\texttt{light} \equiv \texttt{up}(\texttt{s}_1)$. This state constraint, so the argument went, gives rise to indirect effects, e.g., the bulb is expected to light up once $\texttt{s}_1$ is switched into the upper position. Taking this as eternal truth is appropriate for an idealistic version of the circuit. But the expectation that the light can be switched on in this way in a *real* circuit depends on additional, tacitly presupposed conditions, e.g., the bulb must not be broken, the battery not be malfunctioning, and the wiring needs to be in order. If the belief is held that all of these conditions are usually satisfied, then we face a situation similar to the Qualification Problem. Namely, abnormal circumstances, now denying the universality of a state constraint and, hence, the occurrence of the indirect effects it is supposed to trigger, need to be assumed away—but again only to a reasonable extent.

The remainder of this chapter is devoted to this problem of accounting for *exceptions* to state constraints, as we call it. It will turn out that this end merely requires the adoption of the insights gained in the context of the Qualification Problem. This means that rather than newly extending the action theory we finally arrived at, it suffices to pursue a specific design strategy for state constraints that admit exceptions. Treatment of these exceptions, which

M. Thielscher: Challenges for Action Theories, LNAI 1775, pp. 119-124, 2000.

is identical to that of abnormal action disqualifications, is then determined by the model theoretic component of our action theory as it stands.

4.2 Disqualified Causal Relationships

Each state, including the initial one, resulting from the performance of action sequences is supposed to be acceptable, that is, to satisfy all state constraints. This implies that every constraint, as it stands, is universally valid, hence so is the occurrence of indirect effects it triggers. In order to account for possible exceptions to the strict truth of a state constraint, the latter is to be suitably weakened by restricting its range of applicability to normal circumstances. Formally, this is accomplished by rewriting a constraint C as $\neg ab_C \supset C$, where ab_C is a new fluent which, if true, shall indicate that there happens to be an exception to C. Validity of the state constraint is thus confined to 'normal', in a specific sense, states.

Any restriction to the applicability of a state constraint should of course be transmitted to all corresponding causal relationships. If an exception to a constraint occurs, then the indirect effects it usually triggers are no longer expected. The following proposition shows that this comes for free if causal relationships are automatically extracted from the modified state constraints following the guidelines of Section 2.5.

Proposition 4.2.1. *Let $\mathcal{E}$ and $\mathcal{F}$ be sets of entities and fluent names, respectively. Furthermore, let C be a variable-free state constraint and $\mathcal{I}$ some influence information. If $\mathcal{R}$ and $\mathcal{R}'$ are the output of the generation procedure depicted in Fig. 2.6 on input $(C, \mathcal{I})$ and $(\neg ab_C \supset C, \mathcal{I})$, respectively, then*

$$\varepsilon \ \underline{\text{causes}}\ \varrho \ \underline{\text{if}}\ \Phi \in \mathcal{R} \quad \textit{implies} \quad \varepsilon \ \underline{\text{causes}}\ \varrho \ \underline{\text{if}}\ \Phi \wedge \neg ab_C \in \mathcal{R}'$$

Proof. Let $C_1 \wedge \ldots \wedge C_n$ be the minimal CNF of C, then the minimal CNF of $\neg ab_C \supset C$ is $(C_1 \vee ab_C) \wedge \ldots \wedge (C_n \vee ab_C)$. Let

$$\neg \ell_j \ \underline{\text{causes}}\ \ell_k \ \underline{\text{if}} \bigwedge_{\substack{l = 1, \ldots, m \\ l \neq j, l \neq k}} \neg \ell_l$$

be any member of $\mathcal{R}$, then there exists some $C_i = \ell_1 \vee \ldots \vee \ell_m$ (where $1 \leq i \leq n$) such that $(\|\ell_j\|, \|\ell_k\|) \in \mathcal{I}$ and $1 \leq j \neq k \leq m$. Accordingly, the disjunct $\ell_1 \vee \ldots \vee \ell_m \vee ab_C$ determines the causal relationship

$$\neg \ell_j \ \underline{\text{causes}}\ \ell_k \ \underline{\text{if}} \bigwedge_{\substack{l = 1, \ldots, m \\ l \neq j, l \neq k}} \neg \ell_l \wedge \neg ab_C$$

in $\mathcal{R}'$. *Qed.*

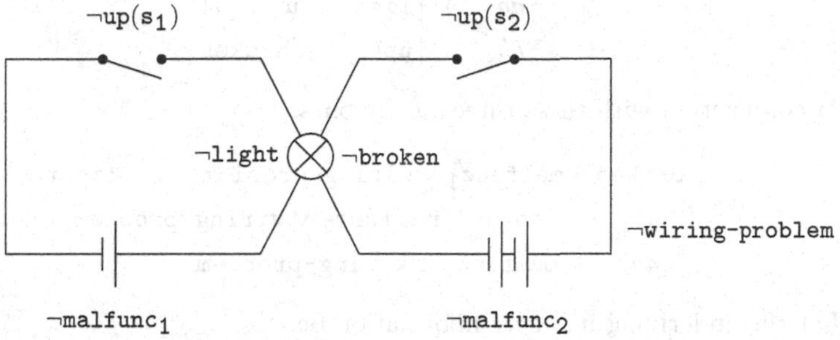

Figure 4.1. In this electric circuit, a light bulb is involved in two otherwise independent sub-circuits. The one on the right hand side includes a battery which is supposed to be so powerful—under normal circumstances—that it destroys the bulb as soon as switch $\mathtt{s}_2$ gets closed. Normally, the two batteries, the wiring, and the light bulb are in order.

It is straightforward to verify that this result is transferable to the generalizations of the algorithm of Fig. 2.6 developed in the section on influence information (Section 2.5).

It is clear that the purpose of introducing a notion of abnormality into state constraints is to assume normal circumstances by default. In the context of the Qualification Problem it proved vital, to this end, to account for the fact that abnormalities may be caused by the performance of actions. It should not come as a surprise if something similar is observed in case of qualified ramifications, and indeed the following scenario shows how exceptions to state constraints can wilfully be brought about. Global minimization of abnormalities is therefore inappropriate as regards state constraints and indirect effects, too.

Example 4.2.1. Suppose the light bulb is involved in a second, otherwise independent sub-circuit consisting of its own battery and switch, see Fig. 4.1. Suppose further that the new battery, without an intermediate resistor, is too powerful for our light bulb so that the latter gets broken as soon as current flows through the rightmost sub-circuit.

As a model of this circuit, let $\mathcal{D}$ be the ramification domain consisting of entities $\mathtt{s}_1$ and $\mathtt{s}_2$; fluent names $\mathtt{up}^1$, $\mathtt{light}^0$, $\mathtt{broken}^0$, $\mathtt{wiring\text{-}problem}^0$, $\mathtt{malfunc}_1{}^0$, $\mathtt{malfunc}_2{}^0$, $ab_1{}^0$, and $ab_2{}^0$; and action name $\mathtt{toggle}^0$ along with the action laws

$$\begin{array}{llll} \mathtt{toggle}(x) & \underline{\text{transforms}} & \{\mathtt{up}(x)\} & \underline{\text{into}} \quad \{\neg\mathtt{up}(x)\} \\ \mathtt{toggle}(x) & \underline{\text{transforms}} & \{\neg\mathtt{up}(x)\} & \underline{\text{into}} \quad \{\mathtt{up}(x)\} \end{array}$$

The set of state constraints shall divide into two non-steady, namely,

$$\neg ab_1 \supset [\mathtt{light} \equiv \mathtt{up(s_1)}]$$
$$\neg ab_2 \supset [\mathtt{up(s_2)} \supset \mathtt{broken}]$$

in conjunction with these three steady ones:

$$\mathtt{broken} \vee \mathtt{malfunc_1} \vee \mathtt{wiring\text{-}problem} \supset \neg\mathtt{light}$$
$$ab_1 \equiv \mathtt{broken} \vee \mathtt{malfunc_1} \vee \mathtt{wiring\text{-}problem}$$
$$ab_2 \equiv \mathtt{malfunc_2} \vee \mathtt{wiring\text{-}problem}$$

Let the underlying influence information be

$$\begin{aligned}\mathcal{I} = \{ & (\mathtt{up(s_1)}, \mathtt{light}), (\mathtt{broken}, \mathtt{light}), (\mathtt{malfunc_1}, \mathtt{light}),\\ & (\mathtt{wiring\text{-}problem}, \mathtt{light}), (\mathtt{up(s_2)}, \mathtt{broken}), (\mathtt{malfunc_2}, \mathtt{broken}),\\ & (\mathtt{wiring\text{-}problem}, \mathtt{broken}), (\mathtt{broken}, ab_1), (\mathtt{malfunc_1}, ab_1),\\ & (\mathtt{wiring\text{-}problem}, ab_1), (\mathtt{malfunc_2}, ab_2), (\mathtt{wiring\text{-}problem}, ab_2)\}\end{aligned}$$

This determines the following causal relationships, all of which except for the first three are steady:

$$\begin{array}{rlll}
\mathtt{up(s_1)} & \underline{\text{causes}}\ \mathtt{light} & \underline{\text{if}} & \neg ab_1\\
\neg\mathtt{up(s_1)} & \underline{\text{causes}}\ \neg\mathtt{light} & \underline{\text{if}} & \neg ab_1\\
\mathtt{up(s_2)} & \underline{\text{causes}}\ \mathtt{broken} & \underline{\text{if}} & \neg ab_2\\
\mathtt{broken} & \underline{\text{causes}}\ \neg\mathtt{light} & \underline{\text{if}} & \top\\
\mathtt{malfunc_1} & \underline{\text{causes}}\ \neg\mathtt{light} & \underline{\text{if}} & \top\\
\mathtt{wiring\text{-}problem} & \underline{\text{causes}}\ \neg\mathtt{light} & \underline{\text{if}} & \top\\
\mathtt{broken} & \underline{\text{causes}}\ ab_1 & \underline{\text{if}} & \top\\
\mathtt{malfunc_1} & \underline{\text{causes}}\ ab_1 & \underline{\text{if}} & \top\\
\mathtt{wiring\text{-}problem} & \underline{\text{causes}}\ ab_1 & \underline{\text{if}} & \top\\
\neg\mathtt{broken} & \underline{\text{causes}}\ \neg ab_1 & \underline{\text{if}} & \neg\mathtt{malfunc_1} \wedge \neg\mathtt{wiring\text{-}problem}\\
\neg\mathtt{malfunc_1} & \underline{\text{causes}}\ \neg ab_1 & \underline{\text{if}} & \neg\mathtt{broken} \wedge \neg\mathtt{wiring\text{-}problem}\\
\neg\mathtt{wiring\text{-}problem} & \underline{\text{causes}}\ \neg ab_1 & \underline{\text{if}} & \neg\mathtt{broken} \wedge \neg\mathtt{malfunc_1}\\
\mathtt{malfunc_2} & \underline{\text{causes}}\ ab_2 & \underline{\text{if}} & \top\\
\mathtt{wiring\text{-}problem} & \underline{\text{causes}}\ ab_2 & \underline{\text{if}} & \top\\
\neg\mathtt{malfunc_2} & \underline{\text{causes}}\ \neg ab_2 & \underline{\text{if}} & \neg\mathtt{wiring\text{-}problem}\\
\neg\mathtt{wiring\text{-}problem} & \underline{\text{causes}}\ \neg ab_2 & \underline{\text{if}} & \neg\mathtt{malfunc_2}
\end{array}$$

Consider, now, a situation where we only know that both switches are open (as depicted in Fig. 4.1). What would be the predicted outcome of first toggling $\mathtt{s_2}$, followed by $\mathtt{s_1}$? Since nothing hints at a malfunctioning battery or bad wiring, we expect that $\mathtt{up(s_2)} \supset \mathtt{broken}$ be true and, hence, $\mathtt{broken}$ is an indirect effect of closing $\mathtt{s_2}$. Consequently, closing $\mathtt{s_1}$ afterwards should fail to produce light. But globally minimizing abnormality in this scenario does not yield this conclusion. Obviously, some abnormality is inevitable. One minimal model is therefore given by $\neg ab_2$ with regard to the first action, and ab_1 with regard to the second. This reflects the expected course of events. Yet we can just as well assume the first ramification unqualified (i.e., ab_2), which then would avoid the necessity of assuming a disqualification of the following ramification (i.e., $\neg ab_1$). For if the bulb does not break as a consequence of toggling $\mathtt{s}_2$, then light turns on as the usual indirect effect of toggling $\mathtt{s}_1$ afterwards. This gives us a second, unintended model, where the more powerful battery is supposed down and light is on in the end. ■

4.3 Causing Exceptions to State Constraints

Global minimization being inappropriate when assuming away exceptions to state constraints, too, our solution to the Qualification Problem furnishes a ready approach to satisfactorily tackling abnormalities in the new context. Since exceptional circumstances can be brought about as side effect of performing actions, fluents representing these circumstances should be made subject to both ramification and, otherwise, persistence. Being *a priori* unlikely to occur, these fluents are to be minimized initially to the largest reasonable extent, just like the fluents that describe abnormal disqualifications of actions.

To summarize, the amalgamation of the Ramification and Qualification Problem is addressed as follows. Each state constraint C which admits exceptions is replaced by the weaker fluent formula $ab_C \supset C$, with ab_C being a distinct new fluent name. The reading of the modified constraint is that now C holds only under normal circumstances. As for the case of action disqualifications, the fluent ab_C should be engaged in additional state constraints defining conditions for an exception to C. In order to reflect that ab_C indicates an exceptional situation, this fluent is subject to minimization, i.e., belongs to the set of abnormality fluents of the domain at hand.

Example 4.3.1. Let $\mathcal{D}$ be the qualification domain of Example 4.2.1, and let

$$\mathcal{F}_{ab} = \{\mathtt{malfunc}_1, \mathtt{malfunc}_2, \mathtt{broken}, \mathtt{wiring\text{-}problem}, ab_1, ab_2\}$$

Let Σ be the transition model of $\mathcal{D}$, and suppose $\mathcal{O}$ consists of the observation

$$\neg\mathtt{up}(\mathtt{s}_1) \wedge \neg\mathtt{up}(\mathtt{s}_2) \ \ \underline{\text{after}}\ \ [\,]$$

Since no abnormality needs to be granted, the qualification scenario $(\mathcal{O}, \mathcal{D})$ admits a unique preferred model (Σ, Res) where $\neg\mathtt{up}(\mathtt{s}_1), \neg\mathtt{up}(\mathtt{s}_2) \in Res([\,])$ and $Res([\,]) \cap \mathcal{F}_{ab} = \{\}$. Performing $\mathtt{toggle}(\mathtt{s}_2)$ in $Res([\,])$ has the direct effect $\mathtt{up}(\mathtt{s}_2)$ and the indirect effects $\mathtt{broken}$ and, hence, ab_1 according to the following causal relationships.

$$\texttt{up(s}_2\texttt{)} \ \underline{\text{causes}} \ \texttt{broken} \ \underline{\text{if}} \ \neg ab_2$$
$$\texttt{broken} \ \underline{\text{causes}} \ ab_1 \quad \underline{\text{if}} \ \top$$

From $ab_1 \in Res([\texttt{toggle(s}_2\texttt{)}])$ it follows that toggling $\texttt{s}_1$ in that state fails to produce the otherwise expected indirect effect $\texttt{light}$. Hence, the only acceptable state which can be assigned to $Res([\texttt{toggle(s}_2\texttt{)}, \texttt{toggle(s}_1\texttt{)}])$ is

$$\{ \texttt{up(s}_1\texttt{)}, \texttt{up(s}_2\texttt{)}, \neg\texttt{light}, \\ \texttt{broken}, \neg\texttt{wiring-problem}, \neg\texttt{malfunc}_1, \neg\texttt{malfunc}_2, ab_1, \neg ab_2 \}$$

Consequently, $(\mathcal{O}, \mathcal{D})$ entails the observation

$$\neg\texttt{light} \ \underline{\text{after}} \ [\texttt{toggle(s}_2\texttt{)}, \texttt{toggle(s}_1\texttt{)}]$$

as expected. ■

4.4 Bibliographic Remarks

By the time this book was written the problem of ramifications having exceptions had received little attention in literature, presumably because satisfactory solutions to the Ramification Problem itself had not emerged until very recently. The proposal to combine solutions to both the Ramification and Qualification Problem has been made in [112, 114]. About the only other papers on that topic are [5, 123]. In both of them expressions resembling causal relationships are allowed to be defeasible. The former presents a formal action theory, whose semantics is, however, indirectly defined via a translation into extended logic programs and by appealing to the notion of answer sets [32]. The latter does not go beyond defining a notion of successor state based on minimizing abnormality. In order to successfully cope with qualified ramifications, therefore, the general approach needs to be adopted of minimizing abnormality in the initial state so that persistence and ramification take care of the further evolution.

Acknowledgments

I am indebted to Wolfgang Bibel for his constant support and to Erik Sandewall for his willingness to serve as the official co-reviewer of this book as my "Habilitationsschrift." I have benefited from many discussions on the topic of this book with Sven-Erik Bornscheuer, Patrick Doherty, Charles Elkan, Bertram Fronhöfer, Dov Gabbay, Steffen Hölldobler, Neelakantan Kartha, Vladimir Lifschitz, Norman McCain, Rob Miller, Rolf Nossum, Javier Pinto, Murray Shanahan, and Hudson Turner. I thank Jürgen Giesl, Christoph Herrmann, Jana Köhler, Stuart Russell, and Thomas Stützle for proofreading parts or even the entire book. I am grateful to Hans-Hermann Thielscher, who supplied me with material for many of the pictures, and to Bettina Weser for formatting the manuscript. I owe thanks to the "Verein zur Förderung der deutsch-amerikanischen Zusammenarbeit auf dem Gebiet der Informatik und ihrer Anwendungen" for financing a memorable stay at the International Computer Science Institute in Berkeley, during which large parts of the theory have been developed and most of the book has been written.

References

1. Lennart Åqvist. A new approach to the logical theory of action and causality. In S. Stenlund, editor, *Logical Theory and Semantical Analysis*. Reidel, Dordrecht, 1974.
2. J. B. Amsterdam. Temporal reasoning and narrative conventions. In J. F. Allen, R. Fikes, and E. Sandewall, editors, *Proceedings of the International Conference on Principles of Knowledge Representation and Reasoning (KR)*, pages 15–21, Cambridge, MA, 1991.
3. Andrew B. Baker. Nonmonotonic reasoning in the framework of situation calculus. *Artificial Intelligence*, 49:5–23, 1991.
4. Chitta Baral and Michael Gelfond. Representing concurrent actions in extended logic programming. In R. Bajcsy, editor, *Proceedings of the International Joint Conference on Artificial Intelligence (IJCAI)*, pages 866–871, Chambéry, France, August 1993. Morgan Kaufmann.
5. Chitta Baral and Jorge Lobo. Defeasible specifications in action theories. In M. E. Pollack, editor, *Proceedings of the International Joint Conference on Artificial Intelligence (IJCAI)*, pages 1441–1446, Nagoya, Japan, August 1997. Morgan Kaufmann.
6. Chitta Baral, Michael Gelfond, and Alessandro Provetti. Representing actions: Laws, observations and hypothesis. *Journal of Logic Programming*, 31(1–3):201–243, 1997.
7. Philippe Besnard, R. Quiniou, and P. Quinton. A theorem-prover for a decidable subset of default logic. In *Proceedings of the AAAI National Conference on Artificial Intelligence*, pages 27–30, 1983.
8. Wolfgang Bibel, Luis Fariñas del Cerro, Bertram Fronhöfer, and Andreas Herzig. Plan generation by linear proofs: on semantics. In *Proceedings of the German Workshop on Artificial Intelligence*, pages 49–62. Springer, Informatik Fachberichte 216, 1989.
9. Wolfgang Bibel. A deductive solution for plan generation. *New Generation Computing*, 4:115–132, 1986.
10. Wolfgang Bibel. Let's plan it deductively! *Artificial Intelligence*, 103(1–2):183–208, 1998.
11. Daniel G. Bobrow, editor. *Artificial Intelligence 13. Special Issue on Non-Monotonic Reasoning.* 1980.
12. Sven-Erik Bornscheuer and Michael Thielscher. Explicit and implicit indeterminism: Reasoning about uncertain and contradictory specifications of dynamic systems. *Journal of Logic Programming*, 31(1–3):119–155, 1997.
13. Gerhard Brewka and Thomas Eiter. Prioritizing default logic: Abridged report. In S. Hölldobler, editor, *Intellectics and Computational Logic*. Kluwer Academic, 1999.
14. Gerhard Brewka and Joachim Hertzberg. How to do things with worlds: On formalizing actions and plans. *Journal of Logic and Computation*, 3(5):517–532, 1993.

15. Gerhard Brewka. Adding priorities and specificity to default logic. In C. MacNish, D. Pearce, and L. M. Pereira, editors, *Proceedings of the European Workshop on Logics in AI (JELIA)*, volume 838 of *LNAI*, pages 50–65. Springer, September 1994.
16. Wolfram Büttner. Unification in datastructure multisets. *Journal of Automated Reasoning*, 2:75–88, 1986.
17. Marie-Odile Cordier and Pierre Siegel. A temporal revision model for reasoning about world change. In B. Nebel, C. Rich, and W. Swartout, editors, *Proceedings of the International Conference on Principles of Knowledge Representation and Reasoning (KR)*, pages 732–739, Cambridge, MA, 1992. Morgan Kaufmann.
18. Alvaro del Val and Yoav Shoham. Deriving properties of belief update from theories of action (II). In R. Bajcsy, editor, *Proceedings of the International Joint Conference on Artificial Intelligence (IJCAI)*, pages 732–737, Chambéry, France, August 1993. Morgan Kaufmann.
19. J. P. Delgrande, Torsten Schaub, and W. K. Jackson. Alternative approaches to default logic. *Artificial Intelligence*, 70:167–237, 1994.
20. Marc Denecker and Danny de Schreye. Representing incomplete knowledge in abductive logic programming. *Journal of Logic and Computation*, 5(5):553–577, 1995.
21. Marc Denecker, Daniele Theseider Dupré, and Kristof Van Belleghem. An inductive definition approach to ramifications. *Electronic Transactions on Artificial Intelligence*, 1998. URL: `http://www.ep.liu.se/ea/cis/1998/007/`.
22. Daniel C. Dennet. Cognitive wheels: The frame problem of AI. In C. Hookway, editor, *Minds, Machines, and Evolution: Philosophical Studies*, pages 129–151. Cambridge University Press, 1984.
23. Patrick Doherty, Joakim Gustafsson, Lars Karlsson, and Jonas Kvarnström. TAL: Temporal action logics language specification and tutorial. *Linköping Electronic Articles in Computer and Information Science*, 1998. URL: `http://www.ep.liu.se/ea/cis/1998/015/`.
24. Phan Minh Dung. Representing actions in logic programming and its applications in database updates. In D. S. Warren, editor, *Proceedings of the International Conference on Logic Programming (ICLP)*, pages 222–238, Budapest, June 1993. MIT Press.
25. Kerstin Eder, Steffen Hölldobler, and Michael Thielscher. An abstract machine for reasoning about situations, actions, and causality. In R. Dyckhoff, H. Herre, and P. Schroeder-Heister, editors, *Proceedings of the International Workshop on Extensions of Logic Programming (ELP)*, volume 1050 of *LNAI*, pages 137–151, Leipzig, Germany, March 1996. Springer.
26. Charles Elkan. Reasoning about action in first-order logic. In *Proceedings of the Conference of the Canadian Society for Computational Studies of Intelligence (CSCSI)*, pages 221–227, Vancouver, Canada, May 1992. Morgan Kaufmann.
27. Richard E. Fikes and Nils J. Nilsson. STRIPS: A new approach to the application of theorem proving to problem solving. *Artificial Intelligence*, 2:189–208, 1971.
28. Joseph J. Finger. *Exploiting Constraints in Design Synthesis*. PhD thesis, Stanford University, CA, 1987.
29. Bertram Fronhöfer. *The Action-as-Implication Paradigm*. CS Press München, 1996.
30. Hector Geffner. Causal theories for nonmonotonic reasoning. In *Proceedings of the AAAI National Conference on Artificial Intelligence*, pages 524–530, Boston, MA, 1990.

31. Hector Geffner. *Default Reasoning: Causal and Conditional Theories.* MIT Press, 1992.
32. Michael Gelfond and Vladimir Lifschitz. Classical Negation in Logic Programs and Disjunctive Databases. *New Generation Computing*, 9:365–385, 1991.
33. Michael Gelfond and Vladimir Lifschitz. Representing Actions in Extended Logic Programming. In K. Apt, editor, *Proceedings of the International Joint Conference and Symposium on Logic Programming (IJCSLP)*, pages 559–573, Washington, 1992. MIT Press.
34. Michael Gelfond and Vladimir Lifschitz. Representing action and change by logic programs. *Journal of Logic Programming*, 17:301–321, 1993.
35. Michael Gelfond and Vladimir Lifschitz. Action languages. *Electronic Transactions on Artificial Intelligence*, 1998. URL: `http://www.ep.liu.se/ea/cis/1998/016/`.
36. Matthew L. Ginsberg and David E. Smith. Reasoning about action I: A possible worlds approach. In *Proceedings of the Workshop on the Frame Problem in Artificial Intelligence*, Lawrence, KS, 1987.
37. Matthew L. Ginsberg and David E. Smith. Reasoning about action I: A possible worlds approach. *Artificial Intelligence*, 35:165–195, 1988.
38. Matthew L. Ginsberg and David E. Smith. Reasoning about action II: The qualification problem. *Artificial Intelligence*, 35:311–342, 1988.
39. Jean-Yves Girard. Linear Logic. *Journal of Theoretical Computer Science*, 50(1):1–102, 1987.
40. Enrico Giunchiglia, G. Neelakantan Kartha, and Vladimir Lifschitz. Actions with indirect effects (extended abstract). In C. Boutilier and M. Goldszmidt, editors, *Extending Theories of Actions: Formal Theory and Practical Applications*, volume SS–95–07 of *AAAI Spring Symposia*, pages 80–85, Stanford University, March 1995. AAAI Press.
41. Enrico Giunchiglia. Determining ramifications in the situation calculus. In L. C. Aiello, J. Doyle, and S. Shapiro, editors, *Proceedings of the International Conference on Principles of Knowledge Representation and Reasoning (KR)*, pages 76–86, Cambridge, MA, November 1996. Morgan Kaufmann.
42. Moises Goldszmidt and Judea Pearl. Qualitative probabilities for default reasoning, belief revision, and causal modeling. *Artificial Intelligence*, 24(1–2):57–112, 1996.
43. Cordell Green. Application of theorem proving to problem solving. In *Proceedings of the International Joint Conference on Artificial Intelligence (IJCAI)*, pages 219–239, Los Altos, CA, 1969. Morgan Kaufmann.
44. Gerd Große, Steffen Hölldobler, Josef Schneeberger, Ute Sigmund, and Michael Thielscher. Equational logic programming, actions, and change. In K. Apt, editor, *Proceedings of the International Joint Conference and Symposium on Logic Programming (IJCSLP)*, pages 177–191, Washington, 1992. MIT Press.
45. Gerd Große, Steffen Hölldobler, and Josef Schneeberger. Linear Deductive Planning. *Journal of Logic and Computation*, 6(2):233–262, 1996.
46. R. V. Guha and Douglas B. Lenat. Cyc: A midterm report. *AI magazine*, Fall:32–59, 1990.
47. Joakim Gustafsson and Patrick Doherty. Embracing occlusion in specifying the indirect effects of actions. In L. C. Aiello, J. Doyle, and S. Shapiro, editors, *Proceedings of the International Conference on Principles of Knowledge Representation and Reasoning (KR)*, pages 87–98, Cambridge, MA, November 1996. Morgan Kaufmann.
48. Joakim Gustafsson. *Extending Temporal Action Logic for Ramification and Concurrency*, Thesis No 719 of *Linköping Studies in Science and Technology*. 1998.

49. Steve Hanks and Drew McDermott. Nonmonotonic logic and temporal projection. *Artificial Intelligence*, 33(3):379–412, 1987.
50. Christoph S. Herrmann and Michael Thielscher. Reasoning about continuous processes. In B. Clancey and D. Weld, editors, *Proceedings of the AAAI National Conference on Artificial Intelligence*, pages 639–644, Portland, OR, August 1996. MIT Press.
51. Steffen Hölldobler and Josef Schneeberger. A new deductive approach to planning. *New Generation Computing*, 8:225–244, 1990.
52. Steffen Hölldobler and Hans-Peter Störr. Complex plans in the fluent calculus. In S. Hölldobler, editor, *Intellectics and Computational Logic*. Kluwer Academic, 1999.
53. Steffen Hölldobler and Michael Thielscher. Computing change and specificity with equational logic programs. *Annals of Mathematics and Artificial Intelligence*, 14(1):99–133, 1995.
54. Joxan Jaffar, Jean-Louis Lassez, and Michael J. Maher. A theory of complete logic programs with equality. *Journal of Logic Programming*, 1(3):211–223, 1984.
55. Antonis Kakas and Rob Miller. A simple declarative language for describing narratives with actions. *Journal of Logic Programming*, 31(1–3):157–200, 1997.
56. Deepak Kapur and Paliath Narendran. NP-completeness of the set unification and matching-problems. In J. H. Siekmann, editor, *Proceedings of the International Conference on Automated Deduction (CADE)*, volume 230 of *LNCS*, pages 489–495, Oxford, July 1986. Springer.
57. G. Neelakantan Kartha and Vladimir Lifschitz. Actions with indirect effects. In J. Doyle, E. Sandewall, and P. Torasso, editors, *Proceedings of the International Conference on Principles of Knowledge Representation and Reasoning (KR)*, pages 341–350, Bonn, Germany, May 1994. Morgan Kaufmann.
58. G. Neelakantan Kartha. Soundness and completeness theorems for three formalizations of actions. In R. Bajcsy, editor, *Proceedings of the International Joint Conference on Artificial Intelligence (IJCAI)*, pages 724–729, Chambéry, France, August 1993. Morgan Kaufmann.
59. G. Neelakantan Kartha. Two counterexamples related to Baker's approach to the frame problem. *Artificial Intelligence*, 69(1–2):379–391, 1994.
60. Henry Kautz. The logic of persistence. In *Proceedings of the AAAI National Conference on Artificial Intelligence*, pages 401–405, Philadelphia, PA, August 1986.
61. Robert Kowalski and M. Sergot. A logic based calculus of events. *New Generation Computing*, 4:67–95, 1986.
62. Henry E. Kyburg, editor. *Probability and Inductive Logic*. Macmillan, 1970.
63. A. R. Lacey, editor. *A Dictionary of Philosophy*. Routledge, 1996. (3rd edition).
64. Vladimir Lifschitz. Formal theories of action (preliminary report). In J. McDermott, editor, *Proceedings of the International Joint Conference on Artificial Intelligence (IJCAI)*, pages 966–972, Milan, Italy, August 1987. Morgan Kaufmann.
65. Vladimir Lifschitz. Frames in the space of situations. *Artificial Intelligence*, 46:365–376, 1990.
66. Vladimir Lifschitz. Towards a metatheory of action. In J. F. Allen, R. Fikes, and E. Sandewall, editors, *Proceedings of the International Conference on Principles of Knowledge Representation and Reasoning (KR)*, pages 376–386, Cambridge, MA, 1991.

67. Vladimir Lifschitz. Restricted monotonicity. In *Proceedings of the AAAI National Conference on Artificial Intelligence*, pages 432–437, Washington, DC, July 1993.
68. Fangzhen Lin and Ray Reiter. State constraints revisited. *Journal of Logic and Computation*, 4(5):655–678, 1994.
69. Fangzhen Lin. Embracing causality in specifying the indirect effects of actions. In C. S. Mellish, editor, *Proceedings of the International Joint Conference on Artificial Intelligence (IJCAI)*, pages 1985–1991, Montreal, Canada, August 1995. Morgan Kaufmann.
70. Jorge Lobo, Gisela Mendez, and Stuart R. Taylor. Adding knowledge to the action description language $\mathcal{A}$. In B. Kuipers and B. Webber, editors, *Proceedings of the AAAI National Conference on Artificial Intelligence*, pages 454–459, Providence, RI, July 1997. MIT Press.
71. Witold Łukaszewicz and Ewa Madalińska-Bugaj. Reasoning about action and change using Dijkstra's semantics for programming languages: Preliminary report. In C. S. Mellish, editor, *Proceedings of the International Joint Conference on Artificial Intelligence (IJCAI)*, pages 1950–1955, Montreal, Canada, August 1995. Morgan Kaufmann.
72. Marcel Masseron, Christophe Tollu, and Jacqueline Vauzielles. Generating plans in linear logic I. Actions as proofs. *Journal of Theoretical Computer Science*, 113:349–370, 1993.
73. Norman McCain and Hudson Turner. A causal theory of ramifications and qalifications. In C. S. Mellish, editor, *Proceedings of the International Joint Conference on Artificial Intelligence (IJCAI)*, pages 1978–1984, Montreal, Canada, August 1995. Morgan Kaufmann.
74. John McCarthy and Patrick J. Hayes. Some philosophical problems from the standpoint of artificial intelligence. *Machine Intelligence*, 4:463–502, 1969.
75. John McCarthy. Programs with Common Sense. In *Proceedings of the Symposium on the Mechanization of Thought Processes*, volume 1, pages 77–84, London, November 1958. (Reprinted in: [79]).
76. John McCarthy. Epistemological problems of artificial intelligence. In *Proceedings of the International Joint Conference on Artificial Intelligence (IJCAI)*, pages 1038–1044, Cambridge, MA, 1977. MIT Press.
77. John McCarthy. Circumscription—a form of non-monotonic reasoning. *Artificial Intelligence*, 13:27–39, 1980.
78. John McCarthy. Applications of circumscription to formalizing common-sense knowledge. *Artificial Intelligence*, 28:89–116, 1986.
79. John McCarthy. *Formalizing Common Sense.* Ablex, 1990. (Edited by V. Lifschitz).
80. Leszek Pacholski and Andreas Podelski. Set constraints: a pearl in research on constraints. In G. Smolka, editor, *Proceedings of the International Conference on Constraint Programming (CP)*, volume 1330 of *LNCS*, pages 549–561. Springer, 1997.
81. Judea Pearl. *Probabilistic Reasoning in Intelligent Systems: Networks of Plausible Inference.* Morgan Kaufmann, San Mateo, CA, 1988.
82. Judea Pearl. Graphical models, causality, and intervention. *Statistical Science*, 8(3):266–273, 1993.
83. Judea Pearl. A probabilistic calculus of actions. In R. Lopez de Mantaras and D. Poole, editors, *Proceedings of the Conference on Uncertainty in Artificial Intelligence (UAI)*, pages 454–462, San Mateo, CA, 1994. Morgan Kaufmann.
84. D. Poole. What the lottery paradox tells us about default reasoning. In R. Brachman, H. J. Levesque, and R. Reiter, editors, *Proceedings of the International Conference on Principles of Knowledge Representation and Reasoning (KR)*, pages 333–340, Toronto, Canada, 1989. Morgan Kaufmann.

85. Ray Reiter. A logic for default reasoning. *Artificial Intelligence*, 13:81–132, 1980.
86. Ray Reiter. The frame problem in the situation calculus: A simple solution (sometimes) and a completeness result for goal regression. In V. Lifschitz, editor, *Artificial Intelligence and Mathematical Theory of Computation*, pages 359–380. Academic Press, 1991.
87. Jussi Rintanen. On specificity in default logic. In C. S. Mellish, editor, *Proceedings of the International Joint Conference on Artificial Intelligence (IJCAI)*, pages 1474–1479, Montreal, Canada, August 1995. Morgan Kaufmann.
88. Erik Sandewall. Systematic assessment of temporal reasoning methods for use in autonomous systems. In B. Fronhöfer, editor, *Workshop on Reasoning about Action & Change at IJCAI*, pages 21–36, Chambéry, August 1993.
89. Erik Sandewall. *Features and Fluents. The Representation of Knowledge about Dynamical Systems.* Oxford University Press, 1994.
90. Erik Sandewall. The range of applicability of some non-monotonic logics for strict inertia. *Journal of Logic and Computation*, 4(5):581–615, 1994.
91. Erik Sandewall. Reasoning about actions and change with ramification. In *Computer Science Today*, volume 1000 of *LNCS*. Springer, 1995.
92. Erik Sandewall. Systematic comparison of approaches to ramification using restricted minimization of change. Technical Report LiTH-IDA-R-95-15, Department of Computer Science, Linköping University, Sweden, 1995.
93. Erik Sandewall. Assessments of ramification methods that use static domain constraints. In L. C. Aiello, J. Doyle, and S. Shapiro, editors, *Proceedings of the International Conference on Principles of Knowledge Representation and Reasoning (KR)*, pages 99–110, Cambridge, MA, November 1996. Morgan Kaufmann.
94. Erik Sandewall. Cognitive robotics logic and its metatheory: Features and fluents revisited. *Linköping Electronic Articles in Computer and Information Science*, 1998. URL: `http://www.ep.liu.se/ea/cis/1998/017/`.
95. Erik Sandewall. Logic based modeling of goal-directed behavior. In A. G. Cohn, L. K. Schubert, and S. C. Shapiro, editors, *Proceedings of the International Conference on Principles of Knowledge Representation and Reasoning (KR)*, pages 304–315, Trento, Italy, June 1998. Morgan Kaufmann.
96. Torsten Schaub. A new methodology for query answering in default logics via structure-oriented theorem proving. *Journal of Automated Reasoning*, 15(1):95–165, 1995.
97. C. B. Schwind. A tableau-based theorem prover for a decidable subset of default logic. In M. E. Stickel, editor, *Proceedings of the International Conference on Automated Deduction (CADE)*, volume 449 of *LNAI*, pages 528–542, Kaiserslautern, Germany, July 1990. Springer.
98. Murray Shanahan. Robotics and the common sense informatic situation. In W. Wahlster, editor, *Proceedings of the European Conference on Artificial Intelligence (ECAI)*, pages 684–688, Budapest, Hungary, August 1996. John Wiley.
99. Murray Shanahan. *Solving the Frame Problem: A Mathematical Investigation of the Common Sense Law of Inertia.* MIT Press, 1997.
100. Murray Shanahan. The event calculus explained. *Electronic Transactions on Artificial Intelligence*, 1999. URL: `http://www.ida.liu.se/ext/emtek/post/1999/02/paper.ps`.
101. John C. Shepherdson. SLDNF-resolution with equality. *Journal of Automated Reasoning*, 8:297–306, 1992.
102. Yoav Shoham. *Reasoning about Change.* MIT Press, 1987.

103. Yoav Shoham. Chronological ignorance: Experiments in nonmonotonic temporal reasoning. *Artificial Intelligence*, 36:279–331, 1988.
104. Lynn Andrea Stein and Leora Morgenstern. Motivated action theory: A formal theory of causal reasoning. *Artificial Intelligence*, 71:1–42, 1994.
105. Mark E. Stickel. A unification algorithm for associative commutative functions. *Journal of the ACM*, 28(3):207–274, 1981.
106. Michael Thielscher and Torsten Schaub. Default reasoning by deductive planning. *Journal of Automated Reasoning*, 15(1):1–40, 1995.
107. Michael Thielscher. An analysis of systematic approaches to reasoning about actions and change. In P. Jorrand and V. Sgurev, editors, *International Conference on Artificial Intelligence: Methodology, Systems, Applications (AIMSA)*, pages 195–204, Sofia, Bulgaria, September 1994. World Scientific.
108. Michael Thielscher. Representing actions in equational logic programming. In P. Van Hentenryck, editor, *Proceedings of the International Conference on Logic Programming (ICLP)*, pages 207–224, Santa Margherita Ligure, Italy, June 1994. MIT Press.
109. Michael Thielscher. Computing ramifications by postprocessing. In C. S. Mellish, editor, *Proceedings of the International Joint Conference on Artificial Intelligence (IJCAI)*, pages 1994–2000, Montreal, Canada, August 1995. Morgan Kaufmann.
110. Michael Thielscher. Causality and the qualification problem. In L. C. Aiello, J. Doyle, and S. Shapiro, editors, *Proceedings of the International Conference on Principles of Knowledge Representation and Reasoning (KR)*, pages 51–62, Cambridge, MA, November 1996. Morgan Kaufmann.
111. Michael Thielscher. On the completeness of SLDENF-resolution. *Journal of Automated Reasoning*, 17(2):199–214, 1996.
112. Michael Thielscher. Qualified ramifications. In B. Kuipers and B. Webber, editors, *Proceedings of the AAAI National Conference on Artificial Intelligence*, pages 466–471, Providence, RI, July 1997. MIT Press.
113. Michael Thielscher. Ramification and causality. *Artificial Intelligence*, 89(1–2):317–364, 1997.
114. Michael Thielscher. A theory of dynamic diagnosis. *Electronic Transactions on Artificial Intelligence*, 1(4):73–104, 1997. URL: `http://www.ep.liu.se/ea/cis/1998/014/`.
115. Michael Thielscher. How (not) to minimize events. In A. G. Cohn, L. K. Schubert, and S. C. Shapiro, editors, *Proceedings of the International Conference on Principles of Knowledge Representation and Reasoning (KR)*, pages 60–71, Trento, Italy, June 1998. Morgan Kaufmann.
116. Michael Thielscher. Reasoning about actions: Steady versus stabilizing state constraints. *Artificial Intelligence*, 104:339–355, 1998.
117. Michael Thielscher. Continuous processes in the fluent calculus. In *Hybrid Systems and AI: Modeling and Control of Discrete + Continuous Systems*, volume SS–99–05 of *AAAI Spring Symposia*, pages 186–191, Stanford University, March 1999. AAAI Press.
118. Michael Thielscher. From Situation Calculus to Fluent Calculus: State update axioms as a solution to the inferential frame problem. *Artificial Intelligence*, 1999.
119. Michael Thielscher. Nondeterministic actions in the fluent calculus: Disjunctive state update axioms. In S. Hölldobler, editor, *Intellectics and Computational Logic*. Kluwer Academic, 1999.
120. Georg H. von Wright. *Norm and Action*. London, 1963.

121. Marianne Winslett. Reasoning about action using a possible models approach. In *Proceedings of the AAAI National Conference on Artificial Intelligence*, pages 89–93, Saint Paul, MN, August 1988.
122. Choong-Ho Yi. Towards the assessment of logics for concurrent actions. In D. M. Gabbay, editor, *Proceedings of the International Conference on Formal and Applied Practical Reasoning (FAPR)*, volume 1085 of *LNAI*, pages 679–690, Bonn, Germany, June 1996. Springer.
123. Yan Zhang. Compiling causality into action theories. In *Proceedings of the Symposium on Logical Formalizations of Commonsense Reasoning*, pages 263–270. Stanford, CA, January 1996.

Index

Lecture Notes in Artificial Intelligence (LNAI)

Lecture Notes in Computer Science

Vol. 1762: K.-D. Schewe, B. Thalheim (Eds.), Foundations of Information and Knowledge Systems. Proceedings, 2000. X, 305 pages. 2000.

Vol. 1763: J. Akiyama, M. Kano, M. Urabe (Eds.), Discrete and Computational Geometry. Proceedings, 1998. VIII, 333 pages. 2000.

Vol. 1764: H. Ehrig, G. Engels, H.-J. Kreowski, G. Rozenberg (Eds.), Theory and Application to Graph Transformations. Proceedings, 1998. IX, 490 pages. 2000.

Vol. 1765: T. Ishida, K. Isbister (Eds.), Digital Cities. IX, 444 pages. 2000.

Vol. 1767: G. Bongiovanni, G. Gambosi, R. Petreschi (Eds.), Algorithms and Complexity. Proceedings, 2000. VIII, 317 pages. 2000.

Vol. 1768: A. Pfitzmann (Ed.), Information Hiding. Proceedings, 1999. IX, 492 pages. 2000.

Vol. 1769: G. Haring, C. Lindemann, M. Reiser (Eds.), Performance Evaluation: Origins and Directions. X, 529 pages. 2000.

Vol. 1770: H. Reichel, S. Tison (Eds.), STACS 2000. Proceedings, 2000. XIV, 662 pages. 2000.

Vol. 1771: P. Lambrix, Part-Whole Reasoning in an Object-Centered Framework. XII, 195 pages. 2000. (Subseries LNAI).

Vol. 1772: M. Beetz, Concurrent Reactive Plans. XVI, 213 pages. 2000. (Subseries LNAI).

Vol. 1773: G. Saake, K. Schwarz, C. Türker (Eds.), Transactions and Database Dynamics. Proceedings, 1999. VIII, 247 pages. 2000.

Vol. 1774: J. Delgado, G.D. Stamoulis, A. Mullery, D. Prevedourou, K. Start (Eds.), Telecommunications and IT Convergence Towards Service E-volution. Proceedings, 2000. XIII, 350 pages. 2000.

Vol. 1775: M. Thielscher, Challenges for Action Theories. XIII, 138 pages. 2000. (Subseries LNAI).

Vol. 1776: G.H. Gonnet, D. Panario, A. Viola (Eds.), LATIN 2000: Theoretical Informatics. Proceedings, 2000. XIV, 484 pages. 2000.

Vol. 1777: C. Zaniolo, P.C. Lockemann, M.H. Scholl, T. Grust (Eds.), Advances in Database Technology – EDBT 2000. Proceedings, 2000. XII, 540 pages. 2000.

Vol. 1778: S. Wermter, R. Sun (Eds.), Hybrid Neural Systems. IX, 403 pages. 2000. (Subseries LNAI).

Vol. 1780: R. Conradi (Ed.), Software Process Technology. Proceedings, 2000. IX, 249 pages. 2000.

Vol. 1781: D.A. Watt (Ed.), Compiler Construction. Proceedings, 2000. X, 295 pages. 2000.

Vol. 1782: G. Smolka (Ed.), Programming Languages and Systems. Proceedings, 2000. XIII, 429 pages. 2000.

Vol. 1783: T. Maibaum (Ed.), Fundamental Approaches to Software Engineering. Proceedings, 2000. XIII, 375 pages. 2000.

Vol. 1784: J. Tiuryn (Eds.), Foundations of Software Science and Computation Structures. Proceedings, 2000. X, 391 pages. 2000.

Vol. 1785: S. Graf, M. Schwartzbach (Eds.), Tools and Algorithms for the Construction and Analysis of Systems. Proceedings, 2000. XIV, 552 pages. 2000.

Vol. 1786: B.H. Haverkort, H.C. Bohnenkamp, C.U. Smith (Eds.), Computer Performance Evaluation. Proceedings, 2000. XIV, 383 pages. 2000.

Vol. 1787: J. Song (Ed.), Indormation Security and Cryptology – ICISC'99. Proceedings, 1999. XI, 279 pages. 2000.

Vol. 1790: N. Lynch, B.H. Krogh (Eds.), Hybrid Systems: Computation and Control. Proceedings, 2000. XII, 465 pages. 2000.

Vol. 1792: E. Lamma, P. Mello (Eds.), AI*IA 99: Advances in Artificial Intelligence. Proceedings, 1999. XI, 392 pages. 2000. (Subseries LNAI).

Vol. 1793: O. Cairo, L.E. Sucar, F.J. Cantu (Eds.), MICAI 2000: Advances in Artificial Intelligence. Proceedings, 2000. XIV, 750 pages. 2000. (Subseries LNAI).

Vol. 1794: H. Kirchner, C. Ringeissen (Eds.), Frontiers of Combining Systems. Proceedings, 2000. X, 291 pages. 2000. (Subseries LNAI).

Vol. 1795: J. Sventek, G. Coulson (Eds.), Middleware 2000. Proceedings, 2000. XI, 436 pages. 2000.

Vol. 1796: B. Christianson, B. Crispo, J.A. Malcolm, M. Roe (Eds.), Security Protocols. Proceedings, 1999. XII, 229 pages. 2000.

Vol. 1800: J. Rolim et al. (Eds.), Parallel and Distributed Processing. Proceedings, 2000. XXIII, 1311 pages. 2000.

Vol. 1801: J. Miller, A. Thompson, P. Thomson, T.C. Fogarty (Eds.), Evolvable Systems: From Biology to Hardware. Proceedings, 2000. X, 286 pages. 2000.

Vol. 1802: R. Poli, W. Banzhaf, W.B. Langdon, J. Miller, P. Nordin, T.C. Fogarty (Eds.), Genetic Programming. Proceedings, 2000. X, 361 pages. 2000.

Vol. 1803: S. Cagnoni et al. (Eds.), Real-World Applications of Evolutionary Computing. Proceedings, 2000. XII, 396 pages. 2000.

Vol. 1805: T. Terano, H. Liu, A.L.P. Chen (Eds.), Knowledge Discovery and Data Mining. Proceedings, 2000. XIV, 460 pages. 2000. (Subseries LNAI).

Vol. 1806: W. van der Aalst, J. Desel, A. Oberweis (Eds.), Business Process Management. VIII, 391 pages. 2000.